Elettra Marconi

Cape Cod 18.1.2003

# MARCONI
# *My Beloved*

by

Maria Cristina
Marconi

First English-Language Edition
Edited, Enlarged and Updated by
Elettra Marconi

DANTE UNIVERSITY OF AMERICA PRESS
Boston

**Library of Congress Cataloging-in-Publication Data**

Marconi, Maria Cristina.
[Mio marito Guglielmo. English]
My beloved Marconi / by Maria Cristina Marconi. -
- 1st English ed. / edited, enlarged, and updated by
Elettra Marconi.
p. cm.
ISBN 0-937832-36-7 (alk. paper)
1. Marconi, Guglielmo, marchese, 1874-1937.
2. Inventors--Italy Biography.
3. Electric engineers--Italy Biography.
4. Radio--Italy--History.
5. Telegraph, Wireless--Marconi system--History.
I. Marconi, Elettra.
II. Title.
TK5739.M3M27313 1999
621.384'092--dc21
[B] 99-44852
CIP

DANTE UNIVERSITY OF AMERICA PRESS
17 Station Street
PO Box 843 Brookline Village
Boston MA 02447
(Distributed by Branden Publishing Company)

Marchesa Maria Cristina Marconi with her daughter Elettra, 1935. (Photo Eva Barrett.)

# CONTENTS

## 6 Maria Cristina Marconi

*The documents on pages 362-367 with the estate of Marconi's grandson, Guglielmo Giovanelli Marconi.*

*Dedico queste mie memorie degli anni vissuti con il mio marito, Guglielmo Marconi, alla mia diletta figlia Elettra e al caro nipote Guglielmo.*

Maria Cristina Marconi

I dedicate these memories of mine of the intense years that I spent with my adored husband, Guglielmo Marconi, to my wonderful daughter, Elettra, and to my dear grandson, Guglielmo.

# A Tribute To GUGLIELMO MARCONI By Carlo Rubbia

What strikes us most forcibly about Marconi's style and the way in which he carried out his work is his modernity. Nowadays we talk more and more about internationalism: our national frontiers coincide with those of Europe and our stage is the world. We are concerned about the purpose and aims of science which must not just produce knowledge but also contribute to improving our quality of life and have useful repercussions for the community. One hundred years ago the young Marconi was also a pioneer in this sense: his ability to be at the same time scientist, inventor and entrepreneur enabled him to operate without frontiers in an international dimension. This is a characteristic which profoundly differentiates Marconi's work from that of the other great scientists of the time.

Marconi, was only in his early twenties when he brought his first experiments to a successful conclusion. Over twenty years had gone by since the great scientific revolution of electromagnetism, from Faraday to Hertz, together with Maxwell's wonderful synthesis of the phenomenology of electromagnetic waves (1873). In that extraordinary victory of human knowledge, the unification of the electric and magnetic fields in electromagnetism, Italy was conspicuous by her absence. It was not until the end of the 1920s, with Enrico Fermi, that the first real school of theoretical physics at world level was established in our country.

How can we explain this paradox? That is, that the fundamental, very important step of applying this knowledge completely escaped the scientific community of the time and that a young man in his early twenties with just a technical diploma, quite obviously on the periphery of the scientific activity of the day, was the first to realize the incredible opportunities offered by the so-called herzian waves for communicating at a distance without wires. In my opinion this was due to an excessive "rationalization of concepts" on the part of the great majority of the scientific community of the time. Electromagnetic waves were then considered essentially like a form of "light". We only have to think of Hertz's classic experiments demonstrating diffraction, refraction, polarization, etc. From hertzian waves to X-rays, all waves were part of the same phenomenon: "unified" electromagnetic waves. Thus, once this mental framework had been accepted, hertzian waves would not have been able to increase the opportunities for using light to any great degree. After all, light does not climb mountains or cross oceans! Hertz died in 1894, one year before Marconi's discovery while Maxwell had died in 1879. Neither of them was there to see the birth of one of the most important practical consequences of their research: the Radio. For all that, Marconi did not live in a cultural vacuum. He studied electromagnetic waves with Righi, who was then a Professor at Bologna; he made an extremely intelligent use of the technological developments of Calzecchi-Onesti, a fellow-countryman living in Lombardy, the inventor of the "coherer".

The scientific and entrepreneurial community in Italy was not culturally ready to support his invention or understand its importance. The Italian government turned down the exclusive rights to his patent and it was only in England that he found the means and the will to develop his discovery. Can we really say that things have changed in the last hundred years? Subsequently, Marconi was also covered with honours in Italy where he was made a life senator. However, jobs were created elsewhere, for example in Great Britain, where the Marconi Company still exists.

Marconi's great superiority and originality lay in his giving priority to the "empirical" side of scientific research, outside and beyond rational conjecture: precisely what we mean by "natural philosophy" or, alternatively, by the "Galilean spirit. Obviously at that time nobody could rationally foresee the existence in the stratosphere, thanks to solar radiation, of a plasma reflector of medium-long electromagnetic waves which channels their propagation around the earth. They did not understand the properties of the long electromagnetic waves which are refracted in the troposphere and therefore follow the terrestrial curvature. Most scientific interest, polarized by Hertz's experiments, was limited to the metric waves which correctly conform to the behaviour of light.

Fortunately, the sun does not emit an appreciable flow of radio waves, otherwise, as in the case of light, all our transmitters would be blinded by it. Just think, for example, that if we look at the sun through a radio-telescope it looks black, a shadow on the luminous radio radiation emitted by our galaxy, the Milky Way.

So we can say that Marconi was a man who had great intuition but who also had good luck: BUT WITHOUT A LITTLE BIT OF LUCK NO GREAT SCIENTIFIC DISCOVERIES WOULD EXIST. Fleming, for example, was a lucky man: how could he have foreseen that while he was studying the behaviour of molds he would discover penicillin.

Marconi's first invention was the antenna-earth structure which is still well-known today. Many years of study and research, in which Marconi was certainly the pioneer, made it possible to transform the Radio into what it is today. Mention must also be made of the introduction of tuning which made it possible for several stations to transmit simultaneously on different frequencies and the introduction of the triode amplifier, discovered by the American, de Forest in 1907. In my opinion, the amplification of signals was the key to success. Today, for example, less than a century later, man has practically reached the theoretical limit of sensitivity, that is, the maximum that is theoretically possible, limited by the noise due to the thermal vibration of the electrodes in the antenna. Just think for example of the pictures transmitted by Voyager from hundreds of millions of kilometres away (the signal took hours to reach the earth) with a transmission power of little more than a torch battery (a few watts). This shows once more the profound, fundamental technological difference between electromagnetic waves and light. In other words, it is as if one could see the light of a little pocket torch shining from the limits of our solar system!

Another example is the picture of the "infant" universe, that is, when it was only 100,000 years old, preserved by radio radiation which, after having travelled for 15 billion years through empty space, is picked up and reconstructed by the COBE satellite.

The Radio and the applications that are associated with it (television, etc.) have helped to unite our planet and we have learnt to know each other better: There is no doubt that this is an important contribution towards making our world a better one.

I should like to conclude with a hope and a wish. If one day we discover that there is life and INTELLIGENT BEINGS in space, it will certainly be because another extraterrestrial Marconi has discovered the Radio out there too. In fact, it has been scientifically proved that communication between galactic civilizations, hundreds of thousands of light years apart, is possible and I WOULD SAY ONLY POSSIBLE with Radio. Today scientists scan the skies with radio telescopes searching for intelligent signals, emitted both consciously and unconsciously by another faraway civilization. (Remember the beautiful and poetic scene in Spielberg's film: "ET-call Home").

The discovery, thanks to the Radio, that we are not the only Intelligent Beings in the Cosmos would be the supreme example and rightful crowning of the extraordinary potential of the invention and intelligence of our Guglielmo Marconi.

# FOREWORD
# by
# Elettra Marconi

The pages you find before you reflect faithfully all the intense, fascinating life story of my mother Maria Cristina Marconi, née Bezzi Scali, beside my father Guglielmo Marconi, he who "gave a voice to silence" by inventing the Radio.

The readers of this book will not find just a technical, scientific text, referring only to historic events, but above all an absolutely truthful story in which an intelligent woman in love with her husband brings alive for the present day the experiences that characterized her exceptional life. Consequently, what stands out is my mother's great joie de vivre and serene happiness during the whole of her unique life with Guglielmo Marconi; in fact, from the first moment he saw her my father had been struck by the sunny nature, blonde beauty and nobility of spirit of the charming Maria Cristina. His union with his beloved wife lasted beyond his lifetime.

My mother wrote down the events of those years in a diary. After my father's death, in the peaceful moments during the summer when we were at Forte dei Marmi or Sardinia she would gaze at the sea with a thoughtful air and relive the many days she had spent sailing on board the *Elettra*. She used to write down her thoughts and read them to me. Her words reflected the inspiration of moments sometimes far apart from one another but always illuminated by the vivid memory of her husband; they have been left in their original form so as to offer to the reader all the intensity of the feelings which united her to him.

What my mother makes clear in these pages, apart from their personal feelings, is the genius of Guglielmo Marconi, who dedicated his whole life to the good of humanity, inaugurating a new era. With his inventions he wanted to join the continents, helping to bring peace and harmony between nations.

My mother was always very clear and incisive when she spoke out to defend her love for my father and to respect his memory but she never made any public declarations and very rarely gave interviews because this conflicted with her extremely reserved character. Initially she wanted to leave these memoirs to me as a private memento of my father. As time went by, however, having read inaccurate and sometimes completely untrue information about Guglielmo Marconi, both as a man and as a scientist, she realised that it was necessary and important to let the world know the truth. When she decided to publish her memoirs she also wanted to add, as historic evidence, some of the letters that her husband wrote to her before and after their marriage.

In accordance with my mother's wishes and to make these pages closer to the young people who want to know and to their elders who remember my father's greatness, for the first time since the death of Guglielmo Marconi I therefore submit to the admiration of the reader the letters that the Genius of the Radio wrote to my mother. She kept them all her life as tokens of an immense, unique, radiant love.

Sadly, she did not live to see the accomplishment of her wishes; she died in fact in July, 1994 (the same month in which my father died in 1937), after 57 years of widowhood dedicated entirely to perpetuating the memory of her beloved husband. Today, as the supreme culmination of their love, my parents lie together in the crypt of Villa Griffone at Sasso Marconi (Bologna), the same country house where my father carried out his first historic experiment which saw the birth of the Radio in the spring of 1895.

My feeling is one of deep gratitude to my mother who kept her diary and wrote her memoirs so faithfully, together with profound emotion for the wonderful and sometimes dramatic experiences of my parents.

After her marriage to my father, my mother grew to know and love Britain; in fact, my father had spent many years there and he transmitted his high opinion of the British to his wife. He was delighted to see how his many British friends took Maria Cristina to their hearts, while she in her turn always spoke of these friends with affection and admiration.

This book, which I present with emotion to its readers, shows in a new and human light the greatness of two exceptional people: my father and my mother.

Rome, September 1995

# INTRODUCTION
# by
# Maria Cristina Marconi

It is difficult and moving for me to speak about my beloved Guglielmo. Many books have been written about his immense work and about his life from his adolescence to his death; but they have not always faithfully reflected the truth.

I dedicated my whole life to Guglielmo Marconi, to this man who was unique for his genius, his great sensitivity and his charm; I was always close to him with my love, helping to stimulate his natural joie de vivre and giving him that serenity that was so necessary to him.

We were united by our great desire to help one another and to overcome every difficulty together. It was like this until the day of his death and nobody was able, although some tried, to destroy our union which instead became more profound with every day that passed.

I wish to emphasise to those who read this book that Guglielmo Marconi was a great scientist; his discoveries and inventions have made it possible for the whole world to communicate, overcoming every distance and saving countless human lives. My husband's genius did not show itself only in the extreme technical rigour with which he tenaciously carried out his numerous experiments but was also expressed in his exceptional generosity of spirit. He was a far-sighted man and he had the good of humanity at heart.

For this reason, overcoming my natural reserve, for love of the truth I want to report phrases, eye-witness accounts and his own authentic words so that everyone will know how much love and sensitivity sprang from the heart of the scientist.

The feeling that united us was very profound. Our marriage was blessed by the birth of Elettra, who before she was born he called "the child of our love", saying to me: "If it is a girl, we will call her Elettra, like the yacht where for all these years I have been carrying out my experiments whilst sailing the different seas.

Guglielmo loved our daughter very much. Looking down at her in her cradle, he whispered to me, smiling tenderly: "She is our angel!"

How many moments of really great importance but also of light-heartedness and harmony we spent together: his great discoveries, the journeys in far-off countries, meeting many people who today are part of history, being received with the magnificent welcome usually reserved for royalty and heads of state, welcomed with open arms by huge crowds applauding our arrival. All these events, with full details, I have described in these pages.

The happy years went by but at a certain moment we both began to realise that the Lord was going to part us after He had willed that we should love each other so deeply. In the last days of Guglielmo's life, before his sudden death from a heart-attack, he looked deep into my eyes after embracing me tenderly and said: "You will have Elettra and she will remind you of me".

Dearest Guglielmo, you were right. You did indeed leave me a living reminder of yourself: our beloved Elettra.

I should like, simply and spontaneously, to write down the truth that comes from my heart and which our daughter Elettra can read over again and keep as a record of her great father and as a testimony of the faithfulness of spirit and mind which her mother always cherished and kept alive in her heart for her husband Guglielmo Marconi even after his death.

Rome, June 1993.

# MY LIFE WITH GUGLIELMO

On 20th July, 1962, the 25th anniversary of my beloved Guglielmo's death, my daughter Elettra and I attended a solemn commemorative Holy Mass in the Cathedral of San Pietro in Bologna, celebrated by Cardinal Giacomo Lercaro the Archbishop of the city in the presence of many dignitaries and a crowd of citizens of Bologna. I was deeply moved and Elettra and I felt very close to our dearest Guglielmo.

Whenever Guglielmo and I went into the cathedral he always wanted to stand with me for a moment or two in front of the font. As I looked at it that day I remembered the words my husband had said to me with a pensive smile on his face: "Look, that is where I was baptised when I was a newborn baby". Although his father Giuseppe Marconi was a Catholic, Guglielmo had been brought up in the Anglican faith by his mother Annie Jameson Marconi who belonged to the Anglican High Church which in the Church of England is the nearest to the Catholic faith.

While Elettra and I were deep in prayer, each one immersed in her own memories, I saw the whole of my life with Guglielmo in my mind's eye with a clarity which made everything seem vivid and alive. I remembered his face, his smile, our first meeting, his honest proud character and all those qualities which made us understand each other perfectly right from the start. I thought of the faith that by God's grace I helped to light and keep burning more and more brightly in his soul which was already so full of God-given greatness and goodness. Finally my mind went back with love and tenderness to our wedding day.

My first meeting with Guglielmo Marconi, by which time he was already free from his first marriage, took place in 1925 during a party on board his yacht the *Elettra* which was anchored at Viareggio. I was with some friends from Florence and Rome among whom were the Duchess Ravaschieri and her daughter Ornella. I was introduced to Marconi who looked at me with interest and gave me a particularly charming smile. As for myself I was immediately fascinated by his great personality and charm and his rather English elegance. I was struck by his aura of genius and I realized that here was an exceptional man. The look in his eyes told me that the first spark of the great love which would eventually unite us for the rest of our lives had already been lit. I was wearing a long red velvet evening dress with a red rose pinned to one shoulder. What a wonderful evening it was.

Not long afterwards Guglielmo and I met again in Rome. I was a close friend of Maria Cristina del Drago, who gave afternoon parties in her palace in Via Quattro Fontane. Her mother Princess Elika was a very charming woman who always invited a select group of friends to those events; the atmosphere was always pleasant and the conversation amusing. After her death her son and daughter Rodolfo and Maria Cristina kept up their mother's custom and entertained in the same style.

Guglielmo Marconi was made a Senator of the Kingdom of Italy when he was just forty years old, the minimum age allowed by the law. He came to Rome from time to time in order to fulfill his duties at the Italian Senate and he used to stay for two or three days at the Grand Hotel. He was always glad of the opportunity to visit his friends Prince and Princess del Drago and it was in the pleasant atmosphere of one of their parties that I met Guglielmo again. During the afternoon I realised that his gaze was fixed upon me and this made me very happy. I was struck at once by the look in his eyes: it was penetrating, compelling and yet full of humanity and feeling. He possessed a mysterious force of attraction which drew people to him and a marvelous sense of humour; everyone enjoyed his company. There was a kind of electric current between us right from the beginning and we understood each other perfectly. As time went by we fell deeply in love and we knew that we wanted to spend the rest of our lives together. Guglielmo often said to me that apart from his mother I was the only person who understood him and he felt truly happy only when he was with me. It was no sacrifice for me to dedicate myself to him completely and I did my best to help him as much as I could. I wanted to transmit my own "joie de vivre" to him.

Guglielmo loved beauty: everything that was beautiful in nature, in man, and in art; he admired the works of such geniuses as Dante, Michangelo, Raffaello, Galileo, Shakespeare and Verdi. When we were in Rome his favourite walks were in the gardens of Villa Borghese, the Pincio and the Gianicolo where he enjoyed looking at the view of the eternal city below us. We went to see Tasso's oak-tree in remembrance of the great poet. Although he was a very great scientist, Guglielmo was also very romantic. We had the same tastes and the same kind of character. We felt a profound affinity which kept us close and made our happiness complete. No disagreement ever cast a shadow over our marriage and we lived in perfect harmony until the end of his life. I was always at his side, supporting him when he was worried and comforting him with my love when things went wrong. We faced every problem calmly and sensibly; this attitude was innate in both of us because we had been brought up like this by our parents and to this day I am grateful to them for the moral strength they instilled in us.

Guglielmo had an exceptional personality. He understood me and this is why we were so happy together. His love and gratitude were my reward for all my care. Above all he trusted me completely; he always wanted to know what I thought and listened to my opinions. He was a man with a thousand interests and with his sense of humour he usually found something to make him laugh. He was curious about everything that went on around him and took an interest in all the important events of the day. His comments on every topic were always kind. In fact he wanted everything to turn out for the best and he thought that every question should be considered from an optimistic point of view. He dealt with every difficulty at once as he did not think there was any point in waiting. “Never put things off”, he always said. Guglielmo had many exceptional qualities that I did not find in other men and his sensitivity and intuition made him understand me at once.

I have many letters that Guglielmo wrote to me from London and elsewhere, before and during our engagement, which show his renewed interest in the Catholic faith. They are profound and interesting. Each of them reflects his nobility of mind and his exceptional human and spiritual values. I always remember a phrase in a letter he wrote to me from London on Christmas Eve, 1926. "...Dear Cristina, whatever happens, you have been the Angel of my conversion, of my redemption, an Angel like the one that stopped St. Paul on the road to Damascus, (...)". When we were married he did not write so often because we were always together and he could tell me his thoughts and feelings personally. To have found someone in whom he could confide and with whom he could trust completely was essential to him. I have about two hundred of his letters. They show a youthful spontaneous spirit and intense emotion which he himself was surprised by and which filled him with joy and optimism about his life and work. As for me I felt very proud to be the source of this happiness that he longed for and deserved so much.

Guglielmo wanted to make a thorough study of the teachings of the Roman Catholic Apostolic Church and he found in my sweetness and my ideals the wisdom, serenity and faith in God which he had missed up until then. "I feel like a ship that has found a safe harbour", he said to me one day. I spoke to him about my faith openly and lovingly; I advised him to study the New Testament together with the commentaries of great scholars. I also gave him the famous ascetic book "The Imitation of Christ" which he found particularly interesting. We often read religious works together. Guglielmo was very interested in the texts of St. Paul and St. Augustine whose works and ideas he admired.

The Sacrament of Confirmation was administered by a Bishop in my family's palace in Via Condotti in Rome, where I was born and where I lived with my parents. My father Count Francesco Bezzi Scali stood godfather to Guglielmo. I was present with my mother Anna Sacchetti the daughter of Marquis Urbano Sacchetti and of Princess Beatrice Orsini. Guglielmo was deeply moved because he knew the importance of the decision he was about to take. I had the joy of knowing that Guglielmo now shared my great faith which was to give us strength for the rest of our lives.

After our marriage, he liked to conclude his scientific conferences on his experiments and inventions by stressing that everything he had succeeded in doing with his discoveries had been a gift from God. I helped him to have a wide religious culture which he assimilated quickly and with great interest. He often met my spiritual counsellor, Cardinal Eugenio Pacelli, who had a great respect for him. He knew that Guglielmo's conversion was genuine and that he truly loved me.

During our engagement I introduced Guglielmo to Cardinal Pietro Gasparri, Secretary of State to His Holiness the Pope. On 11th February, 1929 he and Mussolini who was then the Italian Prime Minister signed the Concordat, thus putting an end to the contention between the Italian State and the Church which had been going on since 1870. I kept the correspondence between the Cardinal and my husband and I presented Pope John Paul II with a letter which Guglielmo sent to Cardinal Gasparri, written on board his yacht; engraved on the writing-paper are the flag of the Royal Yacht Squadron and the name, "Elettra". It is now kept in the Vatican Museums.

On his return to the practice of the Catholic faith Marconi asked the Sacra Rota (the ecclesiastic high court) to annul his marriage since his first wife had already remarried. After carefully examining his matrimonial situation the Sacra Rota granted his request and at the beginning of 1927 Guglielmo's first marriage was declared null. The statement of reasons in the judgment, expressly provided for by the code of canon law, was of "invalid consent" (that is that the parties even before they were married had declared themselves ready to separate in the event of the marriage vows being broken). This statement of reasons was proved conclusively by the unanimous declarations made by the witnesses before the ecclesiastic high court and these statements in fact rendered the marriage null.

This made it possible for us to be married in the Basilica of Santa Maria degli Angeli in Rome on 15th June 1927. It was a solemn occasion and the marriage was blessed by our great friend Cardinal Evaristo Lucidi since Cardinal Pacelli was in Berlin because of his diplomatic commitments as Apostolic Nuncio. Guglielmo's witnesses were Prince Clemente del Drago and Prince

Ludovico Spada Potenziani while my witnesses were my uncles Prince Domenico Orsini and the Marquis Guglielmo Guglielmi d'Antognolla. All our friends and relatives and a crowd of Roman citizens were present and when we came out of the church they showered us with flower petals. After the wedding there was a very elegant and intimate luncheon at my parents' house in Via Condotti with our witnesses, relatives and closest friends.

My wedding-dress was of white satin with a long train, made by the well-known dressmaker Ventura and designed by Madame Anna. I wore a diamond tiara and an antique Irish lace veil which had belonged to Guglielmo's mother on my head. One thing which still touches me is that the tiara was designed for me personally by Guglielmo as he did not think that any of the ones he had seen in the various jewellers in London were beautiful enough for me. (*See letter dated 14 May 1927)*

Since the matrimonial concordat between the Italian State and the Holy See was not yet in force in 1927, the civil ceremony had taken place two days before in the Capitol. This was followed by a formal reception at my parents' house with all our relatives, including Guglielmo's brother Alfonso who had come from London for the occasion and many of our friends from the Roman aristocracy, the Vatican and other personalities, including the ambassadors of various countries who admired Guglielmo very much.

# LETTERS WRITTEN BY GUGLIELMO MARCONI TO HIS BELOVED CRISTINA BEFORE THEIR MARRIAGE

# PRESENTATION

This is the first time that I have published these letters written by my father to my mother.

In life, my parents were always very reserved about the deep feeling they shared for one another. I feel a great deal of emotion in revealing my father's state of mind when he wrote these letters to my mother. I have been uncertain, but at the same time, I have felt the duty to let people know the truth about the great love and devotion that existed between them, which lasted all their lives.

From these letters--so intense and alive, there appears evidence that their love never failed, never had ups and downs, and was always constant, always faithful, always total. I repeat: it is the first time that I offer to those of you who have treasured and have been inspired by my father's memory, and now also, of my mother. I am offering you, the readers of these memoires, the chance to penetrate in the private life of my parents.

I can say with pleasure that my father found, in his love for my mother, a new impulse and the challenge to do and to create always more, to research every aspect and detail of his experiments, and to complete with utmost devotion the preparation of his inventions.

He was, without doubt, always reinforced and enlightened by his Faith, and by my mother's love.

My spirit, as his grateful daughter, being now the only one to keep alive the most intimate, clear memories of both my father and my mother, I ask all you, who read these letters, to do it with respect and understanding.

Elettra Marconi

## 26 Maria Cristina Marconi

*Savoy Hotel London, Sunday 27th June, 1926*

*Dear Cristina,*

*Here I am writing to you again!!!*

*I must tell you that there is no power on earth which can change or diminish by so much as a milligram the greatness and beauty of everything that I feel "for one special person".*

*I have not gone on board the Elettra yet because I did not want to risk my absence delaying the proceedings for my case before the Westminster Court by even one hour. I have started working on my scientific experiments again and also all the other matters regarding radiocommunications which I deal with myself. I feel, from the interest you take in my experiments, that you are pleased that I am going on with my work which keeps me in contact with the beauties of nature that you too love so much.*

*I enclose some cuttings of newspaper articles, one of which gives quite a full description of what I am doing and what I still need to do for communications with the different parts of the British Empire. Another is my description of microwaves. They want to celebrate the 30th Anniversary of my first invention here in England too. As you will see there are some newspapers which are going on a bit about the fact that it was England that understood the importance of my invention of wireless telegraphy and exploited my discovery.*

*I must say that I feel rather embarrassed about sending you newspaper articles that say so many nice things about me but I promise it is not the reason I am sending them. It is because you have told me more than once that you would like me to send them to you, so here they are; also so that you can follow what I am doing from afar and know what is being said about me.*

*And then, as you know, I am determined (as soon as I am completely free) to ask you something very serious and very important. The greatest, most important and serious question that a man can ask and a woman can answer in all her life. Probably this is the reason why I feel such a great need to keep in touch with you in spirit.*

*With my kindest regards to your mother, I remain your devoted,*
*Guglielmo Marconi*

*London, 16th February 1927*

*...The delegates of the American, French and German governments are here in London now to discuss many international questions regarding the Radio and also to make agreements for the use of my Beam System in their respective countries...*

*...Work is never ending but if you were with me everything would be different. The only work which still gives me pleasure when I am away from you is the work which brings me into contact with the beauties of nature. It always seems to me that nature is part of you, part of your soul and I feel that it is also part of mine...*

*London, 18th February, 1927*

*...The other evening when I was in my room I heard part of Lucia di Lammermoor from Rome...and listening to that beautiful voice full of harmony and feeling I felt sure that you were listening to it too. Now I read in your letter that you really were!! And so the Radio has joined our sweet sensations! It won't be long now before the television will make it possible for me to see you too but before that is really possible I hope to be with you always, to look into your eyes as it always was and always will be when there is true love. This ancient and eternal method will always be the best, better even than our Radio!...*

*Savoy Hotel, London, 26th February 1927*

*...But at this point not even the sun would make me happy if I didn't know, if I didn't feel that beyond the seas and the mountains are you, my sun, loving me and waiting for me patiently and faithfully!*

*...I am sending you some newspaper cuttings about me and about the Radio. You will laugh! (how I wish I could see you!) when you read the ones that say that your Wizard is the greatest man in the world!...*

*...If it weren't for this Faith, the edifice of my whole existence would collapse...*

*...I find it difficult not to write to you perhaps even twice a day, if only to say to you again and again that I love you...*

*Savoy Hotel, London, 28th February*
*...It is midnight...Here at this moment I feel like a caged lion who bites the bars wanting to attack anyone who makes you suffer. Yes, my darling Cristina, I am ready to do anything to defend you from those who dare to attack you; I long to leave everything here and rush to your side...*
*...I am touched by the sweet messages that your dear Mother has sent me through you. After your Love, I value her affection for me most of all. She has contributed so much to make you the person you are!*

*Savoy Hotel London, 6th March 1927*
*...I have been for a beautiful solitary walk. I so much prefer the country to the city! I thought of you and of how happy I shall be one day when I can always enjoy the beauties of nature with you--you who are also Her sublime creation...*
*...I love to look at the sky and as I look at it I think that you too often look up...*
*...The General Meeting of the Marconi Company is fixed for 15th March which means that I shall not be able to come to Rome until after that date. It is going to be rather a stormy meeting with the two rival factions of shareholders. It will also be embarrassing for me because both sides want me with them...*

*Savoy Hotel London, 8th March 1927*
*...You are the soul mate that I never expected to find and I repeat that I want you to be my sweet guide and help for the improvement and progress of my spirit which belongs to you completely...*
*...I shall have so many things to tell you and explain to you about my plans for voyages on the Elettra and my new inventions...*

*London, 9th March evening, 1927*
*My darling Cristina,*
*I have just received your sweet letter. I must tell you right away that I have already arranged everything so as to be able to leave on the morning of Sunday 20th and God willing I shall arrive in Rome in the evening of Monday 21st. How happy I am! I have always loved our beautiful country but since I discovered you and*

*all that you are I do not think that even Moses could have longed for the Promised Land as much as I long to see Trinità dei Monti, Piazza di Spagna and Via Condotti.*

*How sweet you are to write to me so often but as you know I have been sure for a long time that you are an Angel.*

*I still have many days of hard work ahead of me before I can leave. The problems with the shareholders are on the way to being settled and it seems that they will follow my advice. On the 15th I have to preside over an important meeting where everything will be decided and from then on I will no longer have to stay on here as in the past.*

*These two and a half months (which have been like two and a half years!) away from you must never be repeated.*

*Just now I am in the middle of the tests with far-off Australia which will end, if all goes well, on the 14th. Tonight I am going to have to stay up very late because we are testing the simultaneous telegraphy and telephony.*

*With the Lord's help I am learning little by little how to transmit these waves right across the world over a distance which is twelve times greater than that which separates London from Rome (I was going to say me from you but there is no longer any distance which really separates us. Our Love is even more wonderful than the Radio!).*

*To reach Australia these Radio waves have to cross the whole of Europe, Mesopotamia, Persia, India and Polinesia, overcoming all the obstacles between them and vanquishing storms and all the hostile forces of nature.*

*The more I work with the forces of Nature and sense the Divine good-will towards mankind the more I am brought into contact with the great truth: that everything is ordered by the Lord and Giver of Life and this so-called science I work with is just an expression of the Supreme Will which wants to put human beings in contact with each other to help them improve and have a greater mutual understanding.*

*In this too God has sent you to inspire me to do greater and greater things. You wrote to me about Brasil and our Government's offer to me. I shall only go if I can go there after the blessed day of our marriage; and only if you want to go. As you say, it might be a good idea for our honeymoon which in any case will never end for us. How wonderful it will be to show you the World and to show you to the World!*

*In the next few days I may not be able to write to you much because I still have so many things to do but remember that if I am working a bit too much now it is so that later I can dedicate myself to you completely because I belong to you and to you alone. The work I will do in the future and which you can order will all be done to please you, my Queen.*

*All I want to do for the rest of my life is to be worthy of you, my darling Angel; I will also try to deserve the pride you feel in a Wizard who belongs to you and who wants to live just for you.*

*The immense love that you have inspired would be quite inconceivable for many people; But not for you--you understand it! You know, don't you? You feel it!*

*Your Guglielmo*

*London, 10th March, 1927*

*My Cristina,*

*Another day has gone by; every day that passes brings me closer and closer to the day when finally I shall see you again.*

*I won't write much this evening because I am very sleepy after staying up so late last night. But there will never again be a day when I won't see you or be with you or write to you if you will let me.*

*The tests with Australia which kept me up almost the whole of last night are going very well but I will write more about it another time.*

*I have to chair a difficult meeting of the Marconi Company on the 15th.*

*Crissy my darling, I'm happy!! And why shouldn't I be! I love the most divinely beautiful, the sweetest, kindest, and most adorable girl in the whole world. We have the same tastes, the same deep and sincere religious feelings without which we could never be really happy; feelings which uplift and comfort us and bring us closer and closer together. And then, such a wonderful thing, I have the Divine gift--the Grace of your Love. Your Love for me is and always will be the most sacred thing in my life, a sign that I have really have been undeservedly blessed by God. You are my ideal, the Angel of my redemption--and among all the beautiful and good things you possess you also have a Mother who is another Angel.*

*I adore you,*
*Your Guglielmo*

*London, 13th March 1927, 6 a.m.*

*...I have had to get up before sunrise since I have to go straight away to the country to visit one of my Radio Stations for the last day of the tests with Australia...*

*...But you are my greatest and most beautiful discovery...*

*Savoy Hotel, London, 15th March, 1927*

*My sweet Crissy,*

*It is very late again tonight but the only real joy in the whole day for me is finally to be able to write to you, if only briefly, to tell you again that the only thing that is really important for me, beside which everything else pales, is our Love and our hopes.*

*However, I want to tell you without delay that all went well for me today at the General Meeting of the Marconi Company which approved all my proposals with a huge majority.*

*Tomorrow I hope to write you a longer letter and I won't fail to sent you some newspaper cuttings which describe how things went. All the tests with Australia have also gone well with the complete approval of the British and Australian Governments. Now I have to wish you goodnight, my Angel; take care of yourself, for my sake too. In spite of the thousand and one things I have had to do today (I didn't even have the time to have lunch!) my thoughts have always been with you--because I love you and adore you with every fibre of my being, now and for ever for as long as I live.*

*Your own Guglielmo*

*Savoy Hotel London, 17th March 1927--Midnight*

*...This evening I am doing the final tests on your little apparatus and to tell you the truth this interests me more than all the others because the new Radio will soon be yours, and then it will*

*always be with you (one day with us!). Tomorrow I have to do some interesting tests on the transmission of pictures to Australia. One day I will show you how it is done and how quickly!*

*Perhaps these pictures which travel to the antipodes with the speed of thought will go via Rome in their journey--perhaps across your room without you knowing anything about it for now? When the day comes that we can go on beautiful voyages together in our dear Elettra it will be easier for me to explain to my clever and beloved pupil how these things are done!...*

*...I know how much you love and appreciate the beauties of nature, the expression of the Divine Will in which the eternal ideal values, Truth, Beauty and Goodness are found (you have all three in yourself). The united Harmony of causes and laws forms the Truth; the united Harmony of lines, colours, sounds and ideas forms the Beautiful; while the Harmony of feelings and will forms the Good. As the highest expression of the Supreme Eternal Creator they connect in the human Being and carry him towards the final perfection...*

*London, 19th March 1927*

*My Crissy,*

*I have just received your dear, sweet letter No. 31. Thank you again for all the sweet things you say which always move me and make me happier than I ever dreamed possible. And tomorrow I am on my way to you.*

*I have thought of you so much in these last days of work and worry, loving you and, although we are far apart, feeling close to you in your prayers which I am certain will not fail to be heard and answered because of your goodness and fervour.*

*This letter will reach you after I have arrived in Rome, if all goes well with my journey. It will get to you after I have looked into your eyes, after I have told you again that I love you, perhaps while I am saying it to you. You will never get tired of hearing me saying it, will you?*

*My love for you is great enough to fill the universe, even more than electric waves can do and I wish you could receive it from every corner of the earth!*

*I can tell that it makes you happy to be loved as I love you with all the strength of my heart and my mind and I need to be loved as you love me with your heart of gold and your angelic sweetness.*

*Our love, which I feel has already been blessed by God, helps us both to comprehend the Divine goodness and beauty, it wakes us and spurs us on, it does us good and promises better.*

*Your Wizard*

*Ritz Hotel, Paris, 1st April, 1927.*

*My Crissy,*

*Today is your birthday and I thank God for the Divine Grace which brought you into the World and into my World so that April's sweet and happy dawn is still more beautiful for me now and forever.*

*You came robed in flowers as Nature awoke to hail the Sun which God sent to give light, life and hope to all His creation.*

*This morning I was so happy to receive your dear, sweet letter No. 1. How many nice, dear and wonderful things you manage to say to me in just a few lines!*

*Just think when you are alone that my spirit is always with you; that there is a being who is completely yours, who feels with you, who exists just for you and who only wants to go on living to love and adore you and make you happy for ever. Here in Paris I feel lost.*

*I think about you all the time and everything I see seems temporary and fleeting as if they will only be able to exist and become reality again when you are with me. I can no longer have any joys or interests unless you are there to share them with me. Without you they mean nothing to me any more.*

*Don't think that I am ungrateful for the Divine goodness and benevolence to which I owe so much, to which I owe everything. But God has given me this Eternal and Almighty Love and I feel He has done it for my good and I dare to think for yours. Perhaps more than anything he wanted me to find in you my Angel of redemption and salvation.*

*I have been out a bit today too but it was very cold. Tomorrow I leave for London.*

*At dinner today I met some old Spanish friends again. The Duke and Duchess of Peneranda and the Marquis and Marchioness of Viana. They gave me their best wishes and said they hoped to be able to give me their felicitations soon! The Duchess of Peneranda, who owns the most beautiful pearls in Europe, was wearing a string of them, all as big as cherries!!*

*But to get back to you, my blessed Angel, you who are worth more than all the pearls in the world, I want you to know that I miss you so much and that I need you to make me complete, to live, to become what I really feel I could be.*

*But just thinking of you gives me a feeling of tender sweetness, a light in the darkness, a joie de vivre that uplifts me and keeps me from what would otherwise be a profound sadness.*

*I love and adore you, praying and thanking God over and over again for your Love.*

*Your Wizard*

*London, Sunday 10th April 1927*

*My Cristina,*

*This morning in Westminster Cathedral I heard the Blessing of the Palms by Cardinal Bourne.*

*I longed for my "dear sweet Guide".*

*Holy Week has begun, consecrated to the Passion and Resurrection of our Lord. I think a great deal about the Cross, the symbol and sign of our Faith because everywhere the Cross brings to mind the agony of Jesus for the sins of the whole world.*

*Together with the birth and life of the Redeemer, the Holy Cross always makes me think of the Supreme Sacrifice by which God was made man to offer us Salvation, teaching us that there can be no forgiveness without sacrifice.*

*I want you to know my thoughts about these things too because I feel I have to tell you everything.*

*Tonight I shall pray and keep vigil till late.*

*I love you Cristina. God Bless you always, my Angel.*

*Your Guglielmo*

*Savoy Hotel London, 14th May 1927*

*...Today I ordered the diamond diadem from Chonmet for your darling little head. I hope you will like it. I designed it myself because I couldn't find anything in the whole of London that I really liked! It will be ready and in Italy in time for our Wedding. You tell me that I have good taste in jewellery and I know your exquisite taste in clothes and everything else. Maybe I could be a jeweller and you a dressmaker--but in the same shop!! I am sure that whatever we did together we would always be happy. For the moment however the Radio gives me a lot of work...*
*...Remember that your promise to marry me is the highest and greatest honour that I have ever had in my life and that your Love is the supreme Grace that the Divine goodness and benevolence has granted me...*

*London, 18th May 1927*
*My Crissy--dear and wonderful fiancée--I got up very early this morning and went to the country to see one of my new beam stations where we receive the radio signals from South Africa. Everything was fine but I must confess that these days I find it difficult to concentrate as I should on my work. But I promise you that when you are all mine I hope to do wonderful things.*

*I thought about you all the time--I can't help it because with your noble mind you possess me completely. Meanwhile, the Radio Station I visited which is near the sea has some magnificent woods around it and this morning it seemed that the whole of nature was waking in the hope and glory of spring! I felt it too...and thinking of you, above and beyond everything, I dreamed of when I shall be able to take you to the woods of Fowey--when at nightfall we can be alone with our great Love in the midst of such wonderful nature.*

*I also thought of Victor Hugo's verses and when I came back here I re-read them and I am writing them down for you because they express exactly how I felt when I was in the woods thinking of you and what I thought and felt at Fowey but did not then dare to write to you:*

*Dans les pâles ténèbres des bois*
*La calme et sombre nuit ne fait qu'une prière*

*De toutes les rumeurs de la nuit et du jour*
*Nous, de tous les torments de cette vie amère*
*Nous ne ferons que de l'amour.*

*I must leave you now, my treasure, because I have a thousand things still to do. Till Saturday evening, my darling future bride. I hold you close and I send you millions of...and of love which will always envelop you because I feel that I have enough to fill the whole universe and more!*

*Your Guglielmo*

*Savoy Hotel London 19th May 1927*

*...Your letters are the most beautiful that I have ever read--and they always go straight to my heart. I shall always thank God for your and our Love and I shall be thankful if I can always bring you comfort and joy--because that is what I always want to be able to give you in exchange for the happiness (not of this world) which I feel thanks to you.*

*What you said to me is absolutely true: "I am certain that few mortals in the course of the centuries have loved each other so deeply and completely as we do; knowing, appreciating and valuing each other--these are all beautiful things which complete our Love". And among all the women of all the ages you were, are and always will be my ideal because you are beautiful in heart, body and mind--because you make me feel that I am rising up to reach a paradise on earth with you and one day we will reach the eternal paradise in the skies together...*

*London, 20th May 1927*

*My darling Fiancée,*

*You will receive this after my arrival, after my return to you, after I have seen you again and perhaps held you close, after I have told you and told you again and repeated the greatest and most beautiful truth of my life which is that I love you with a love which seems Divine to me and greater than I believed Love could ever be before you, my Angel of Love, came into my life.*

*And so, by the Grace of God, this really is my last day in London without you. Tomorrow I shall see the English country-*

*side--the sea--France--Paris--the Alps--for the last time before my dream of happiness comes true. Already everything seems more beautiful, more divine, more full of hope and Faith for this life and hereafter!...*

*And I am still dreaming--of you, my dream. What is absolutely certain is that I am yours forever, that I love and worship you, that I am on the threshold of Paradise.*

*Thousands of...with my sweetest and most beautiful thoughts.*
*Your wizard and fiancé*
*Guglielmo*

*Savoy Hotel London, 21st May 1927*

*...soon to make you my darling and beloved bride--the greatest of all the countless blessings that God has given me...*

*...but my Love is so immense and your power to receive it and return it is so infinite that I will never be able to say or write all that I feel to you!...*

*Naples, S.Y. Elettra, Friday*

*...but beyond, through and above all, I always seem to see your radiant beauty and the sweetness of your expression (unique in the world)...*

*...I am leaving almost at once so as to be in Rome in time to see you this evening. I am leaving the Elettra with warm thoughts because I feel that she is going to be our floating home--and I know that you are fond of her too...*

*Grand Hotel, Tuesday night*

*...In all my thoughts and in all my actions you, my wonderful Cristina, are and always will be my inspiration for all that is beautiful, good and great...*

*...you will inspire me to great things, you will lead me towards the paths of heaven; so that together, by the Grace of God, we will reach a happiness that is not of this world...*

*Sunday, Midnight*

*...my divine Cristina, you are and always will be the incarnation and the fulfilment of my great bright dream of beauty,*

*goodness and peace in which the poetry of all the poets unites with the harmony of the most Divine music...*

*Tuesday night*

*...All virtues, like everything great and beautiful, are daughters of the Love that comes from God; the love that has inspired the poets, the food of the generous souls that from their solitude pass on their celestial songs to the last generation, sending me spiritual thoughts from heaven which seem to be made just for you...*

*The Grand Hotel, Rome, Wednesday night*

*...before giving thanks once again to God because you exist, I want to repeat to you that I'm happy--happy in our special and extraordinary Love, happy because the days are passing and bringing us closer to even happier days when I'll no longer have to leave you, when we'll be as one--when you will be my constant comfort and my dear sweet guide -when I'll have the right and the privilege to protect you--and the world will have to respect and admire our Love...*

*The Grand Hotel, Rome, Tuesday night, 7th June, 1927*

*...My Crissy, my sun, beacon and light of my port, of my refuge, of my life's Paradise...*

*Sunday night, 12th June, 1927*

*...I feel that I can only thank God humbly for the immensity of his Grace in giving us to one another and tomorrow I will greet the day with my soul in ecstasy, completely absorbed in you, my Angel of goodness and beauty...*

# LETTERS WRITTEN
# DURING THEIR MARRIAGE

*Crissy my Angel*

*I was so happy to hear on the telephone from Mammà that both you and our Elettruzzi are better. Now, my Crissy, you need to take care of yourself so as to get completely well. You will probably feel much better now that the worst of the cold is over.*

*Everything went well at the Academy this morning but I was sad not to have you there.*

*The King asked after you as we were going upstairs and I told him how sorry you were that you couldn't come because you had a bad cold. He asked me to send you his best wishes and to tell you that nearly everyone has a cold at the moment. All the people I spoke to asked after you.*

*The decoration of the Order of Malta looked really magnificent. But I missed you more than I can say.*

*This morning was a bit empty since there was no letter from you but I'm hoping that there will be one tomorrow. I have meetings tomorrow and Tuesday but on Wednesday I'll do my best to come to you because I can't live without you. Sorry this is in such haste.*

*Millions of kisses and hugs to you my Love; and lots of hugs, but not so hard as to hurt her, to our Elettra.*

*I am so grateful to Mammà and Papà for all they do for us.*

*Your own*
*Guglielmo*

*Elettra, 16th August, 1934*

*Crissy My Darling,*

*Just a couple of lines in haste. We're leaving from Civitavecchia for Messina. Admiral Monaco has come on board and he is going to do the round trip to Venice with me. I sent him a message saying I was leaving today and would like to have his company and he managed to join me here. He sends you his kindest regards.*

*The crossing from Porto Santo Stefano was perfect, the sea absolutely calm so that the portholes could be left wide open. All this makes me miss you because by now everything on board reminds me of you and Elettra.*

*I can't wait to get to Venice so as to join you at Ortisei.*

*Thank you for your telegram which has made me feel happier.*

*I hope that you are all well and that it isn't too cold. I am very busy with my conferences. You don't know what I would give to have you here on board with me, especially knowing that with this weather you wouldn't get tired.*

*With a thousand loving kisses to you and Elettra and best wishes to everyone.*

*Ever your own*

*Guglielmo*

*Hotel Splendid, London, 1935*

*Just to tell you that I'm better, that I've rested a little and that I Adore you more than ever.*

*Guglielmo*

*Rome, 1937*

*Crissy my darling Angel*

*You can't imagine how very sad I am to know that you are ill and not be with you. It's true that there is Mammà but she does so much for Elettra that I wouldn't like her to get overtired looking after you both.*

*Thank you my darling for your adorable letter. How kind you are to write so many interesting and sweet things to me even though you are feeling so weak.*

*I am honoured that Papà has given me his Cross and collar of the Order of Malta as a token of his great affection. It's beautiful.*

*Tomorrow at the Academy reception I'll tell the King and everyone why you couldn't be present.*

*I'm afraid you have influenza and I beg you not to travel until you are completely better.*

*My darling, I Love and Adore you. Your own*

*Guglielmo*

# EXCERPTS FROM GUGLIELMO'S LAST LETTERS AND MESSAGES

*Royal Academy of Italy--the President--8th March, 1937*
*Darling Crissy,*
*...Until this evening on the telephone.*
*Lots of kisses to you and Elettra.*
*Guglielmo*

*Senate of the Realm -29th April, 1937*
*My darling Crissy...*
*...So much love and kisses to you and Elettra.*
*Guglielmo*

*3rd May, 1937*
*My darling Crissy...*
*A thousand kisses to you and Elettra.*
*Until this evening.*
*Guglielmo*

*Royal Academy of Italy--8th July, 1937*
*My darling...*
*...All my love for you and Elettra, darlings. Hoping to see you tomorrow.*
*Guglielmo*

# GUGLIELMO MARCONI'S TRANSMISSION ACROSS THE ATLANTIC CORNWALL TO THE ISLAND OF NEW-FOUNDLAND 12th December, 1901

From the first day of our marriage my life with Guglielmo was spent mostly on board his yacht the *Elettra* sailing in the Mediterranean, the Atlantic and the North Sea. I remember our cruises off the coast of England. We usually set sail from Southampton and sailed along the lovely coast of Cornwall which was one of our favourite places. We spent most of our time on board but sometimes we left the yacht in a harbour or cove for a few hours to go and visit friends who lived in old manors or in splendid country houses.

We often dropped anchor in Poole Harbour, not far from the places where years before Guglielmo had carried out very important experiments and transmissions. We went ashore in the motorboat to the jetty belonging to Lord Montagu. After walking through a beautiful wood we found ourselves in front of a XIII Century Abbey which had been converted into a country house. We were welcomed by Lord Montagu himself, smiling and jovial. We spent happy hours with him and his friends and in the evening we returned to the *Elettra* and continued our voyage.

Guglielmo pointed out a few rare palms growing here and there along the coast. He liked the warm climate of Cornwall which was milder than that of the rest of England. After our stay in those pleasant surroundings we spent months and months in the fogs of London. Although Guglielmo was half English he had an inborn longing for the sun. And yet his scientific work, his experiments and his important responsibilities with the Marconi Company in London meant that for around forty years he spent most of his time in northern climes. It was only after our marriage that he began to return to Italy frequently and feel a closer link with the land of his birth.

When we disembarked from the *Elettra* to visit our friends we would find our very nice Russian chauffeur Bindoff, who had followed our route by road, waiting with the beautiful Rolls Royce to take us wherever we had planned to go. We would drive to join our friends in their stately homes, those unforgettable monuments to past centuries. We often visited Constance Cornwallis-West, the first wife of the Duke of Westminster, who was always delighted to see us and gave us a great welcome. Although she was no longer a young woman she was still beautiful, like her sister the Princess of Pless.

We sometimes went as far as Lizard Head in Cornwall and disembarked to visit the Poldhu Hotel which was on a sheer cliff on a promontory overlooking the ocean. Guglielmo had spent many months there in 1900 and 1901 while he was building the radio-transmitting station of Poldhu right behind the hotel on the cliffs by the seashore. He often spoke to me about that exciting time. He was only twenty-seven years old! His audacious plan was to send the first message across the Atlantic between Cornwall in England and the American continent. The Poldhu radio station had just been completed when the antenna system was destroyed by a violent storm. "I wasn't discouraged", Guglielmo told me with a smile. "It was all for the best because I set to work again at once and put up a new experimental antenna which gave very satisfactory results".

In 1901 he also built a very large radio station on Cape Cod in Massachusetts. He arrived at Cape Cod with his assistants Mr. Kemp and Mr. Vyvyan. First he considered Barnstable but it was too far inland. Then he decided on Highland Light but the local people were suspicious and refused to sell him any land to build his station. Finally he was able to buy a piece of land at South Wellfleet on a high headland of dunes facing the Atlantic. There was nothing between it and the station at Poldhu but the Atlantic

Ocean. He set up his headquarters at a boarding house, the Holbrook House in Wellfleet. The natives of Cape Cod predicted that the circular aerial system of twenty 200 foot masts would be blown down in the first Atlantic gale. They were proved right when the South Wellfleet station was completely destroyed in a violent storm. There was no hope for the time being of transmitting between Poldhu and Cape Cod and Guglielmo decided to move to Newfoundland. He had already spent months sailing along the Atlantic coasts in search of the most suitable promontory for his radio transmissions. He landed at Saint John's, Newfoundland, the nearest point to Poldhu. He was received by the Governor Sir Cavendish Boyle and the Premier Sir Robert Bond.

Guglielmo immediately set up a rudimentary radio-receiving station on a hill near the sea, called "Signal Hill". His two most trusted assistants Mr. Kemp and Mr. Paget had accompanied him from London. A stone-walled cabin panelled inside in wood was put at his disposal. By now it was winter. The building was surrounded by snow and ice but inside it was well-equipped and heated and they were very warmly dressed. Guglielmo had a table in the cabin for his instruments where he spent many hours of the day and night studying and making experiments. The North winds blew, freezing and violent. Guglielmo had suspended the antenna in the air using balloons which were destroyed by the storm. So he thought of holding up the antenna with a kite but this too was blown away. Undismayed, he immediately had another sent up in its place.

He was very young at that time and attempting to do what others had not even dared to imagine. Failure was out of the question. Finally his enthusiasm and determination were rewarded; he heard the signal of the letter "S" for the first time, sent across the ocean from the distant transmitter of Poldhu. That day, 12$^{th}$ December 1901, Guglielmo sat listening in at the Newfoundland radio-station. He had not informed the Press because he did not want any publicity. An assistant at Poldhu in Cornwall was ready to transmit the agreed signal. In between lay the Atlantic, a distance of one thousand eight hundred miles, thought by others to be insurmountable.

It was twelve thirty when Guglielmo heard three brief clicks in rapid succession, faint but clear. He passed the ear-phone to his assistant and asked: "Do you hear anything Mr Kemp?" "Of course!" came the reply. "It is the letter "S". Guglielmo later recorded the following words on a gramophone record: "Then I knew that I had been right. The electro-magnetic waves sent from Poldhu reached the other side of the Atlantic, serenely ignoring the terrestrial curvature which according to some incredulous people should have been an unsurmountable obstacle; I realised then that the day was not far off when I would be able to send complete messages across the continents and the oceans. At that moment long distance radio-telegraphy was born".

When Guglielmo heard the first signal that came from Poldhu, he was filled with an indescribable happiness. He was absolutely certain that he would succeed. He often told me so: "I never had a moment of discouragement or doubt. I was quite ready to stay there in the snow, putting up with all that discomfort for a long time until I finally reached my goal". This was how he always acted; he followed his ideas through with enormous perseverance and determination. I knew how tenacious my husband could be and I never had any doubts that he would succeed in whatever he resolved to do.

Whenever he spoke to me about "Signal Hill" at Saint John's Newfoundland and the moment when he received the sound of the letter "S" across the Atlantic, Guglielmo relived the incredible emotion he had felt. His face lit up and he looked youthful and full of enthusiasm. He realized that he had cancelled the great distances which separated the old world from the new and brought the people of the two continents closer together. He looked to the future. He was perfectly aware of the immense value of his invention and proud of the benefits it would bring to humanity. However, in spite of this, he was modest and altruistic. He was like the great men of the past who made a name for themselves with their works of genius without thought for the personal gain that these would bring them.

Guglielmo told me that the Canadian government immediately understood the importance of his discovery. The Anglo American Telegraph Company had threatened court action over the infringement of its monopoly on all telegraphic business in Newfoundland and rather than challenge the Company's monopoly Marconi decided to withdraw from the colony. The Canadian Minister of Finance, Sir William S. Fielding invited him to continue his experiments in Canada, assuring him of the Canadian government's co-operation and financial support. The Canadians knew that his transmissions across the Atlantic would help their ships and facilitate relatives with England.

The day after the first transmission across the Atlantic the story was on the front page of all the newspapers and Guglielmo was described as "the wizard of space". The "New York Times" wrote: "Marconi's initial success captures our imagination. All men of intellect hope profoundly that the Wireless will soon show that it is not a scientific toy but a system of ordinary and everyday use. Men of science point out the obstacles, obstacles that are generally declared to be insurmountable; but the first triumph is a hope for future conquests".

While many men of science were indeed skeptical and refused to believe that he had actually heard the signal from Poldhu, Professor Michael Pupin at Columbia University believed him and stated: "According to the newspapers I have read, the signals were weak but this is unimportant--the distance has been overcome, now they just have to perfect the transmission equipment. Marconi has definitely proved that the curvature of the earth is not an obstacle for wireless telegraphy. We can only regret the fact that many so-called scientists have tried to take away from him and his assistants the merit and benefits of the work to which they have every right". Elihu Thomson, one of the most important pioneers of electricity in America wrote a letter to the editor of the magazine "Electric World", T. Comerford Martin, affirming his complete faith in Marconi. Many sceptics changed their minds after reading this letter.

Guglielmo had happy memories of his arrival in New York where he received high praise from the Government and Press. Many scientists and personalities congratulated him. A great banquet was given in Guglielmo's honour in the Astor Gallery of the Waldorf Astoria hotel in New York on 13$^{th}$ January, 1902, under the patronage of the American Institute of Electrical

Engineers and organized by "Electric World" and a group of scientists and business men. There were more than three hundred guests at the banquet while a huge crowd gathered on the balconies above the great dining hall. On the wall behind the table of honour hung a plaque with "Marconi" spelt out in electric light bulbs. To the right of the table another plaque said "Poldhu" while on the left was written "St. John's". Between these three sparkling words hung light bulbs which twinkled on and off giving the "S" sign of the Morse Code.

At the end of the banquet Martin read out the letters and telegrams which the organizing committee had received. Edison wrote: "I am sorry not to be present to offer my congratulations to Marconi. I would like to meet that young man who has had the monumental audacity to attempt and to succeed in jumping an electric wave across the Atlantic". When Guglielmo rose to his feet to thank those present he was filled with happiness and emotion. He thanked the American Institution of Electrical Engineers and expressed his "honour to be amongst so many eminent men whose names are familiar in the whole world". He modestly acknowledged what he owed to the scientists who had come before him in the study of electric waves. He spoke about his experiments at Signal Hill and his 7777 (four sevens) patent which he had taken out on 16$^{th}$ April, 1900 and which would make it possible to improve the transmissions, saying: "Thanks to the experiments and improvements we have made, the messages can be read only when the receiver and the transmitter have been tuned". The only mention he made of his problems with the cable companies were the following words: "Laying the underwater cables costs so much that the telegraph companies have to impose very high charges for the service. My system will lower the costs a great deal". He ended his speech with: "I drink to the health of the American Institute of Electrical Engineers".

The dinner was a triumph for Guglielmo and the next day the *New York Times* wrote in its editorial: "Marconi's words were so modest, so lacking in any exaggeration for commercial ends, so generous in recognizing his debt towards the pioneers of research whose path he has followed, so frank in giving credit to the alive and the dead and nevertheless so cautious in giving advance notice of the developments of the work he is carrying out that all those present felt obliged to give Marconi not only the honour of his discovery but also the higher honour which is due to him who puts the truth before every jealousy and professional rivalry. From the laurel crown fashioned for his head he took branches to make garlands also for all his predecessors and colleagues in the study of electric waves and this spontaneous gift has enriched rather than impoverished him".

In a letter to the *The Times* of London Professor Ambrose Fleming wrote: "When one thinks that those dots and dashes are the result of electric wave trains travelling at the speed of light in infinite space, picked up by the thin wire of an antenna, automatically severed and translated by two pieces of apparatus into intelligible messages in different languages, the wonder of it all cannot but strike the imagination".

In Canada, Guglielmo founded the Marconi Wireless Telegraph Company, which was controlled by the Marconi Company of London. He was delighted to hear the news that Thomas Edison had bought shares in his company.

Alexander Graham Bell had become a friend of Guglielmo's and had told him that he could use some land belonging to him on the beach of Beinn Breagh for his new radio station; However, Guglielmo had to refuse the offer, because he realized it was too far inland from the Atlantic to allow him to transmit across the ocean. Guglielmo visited different places along the Canadian coast accompanied by the Canadian personalities of the day. It was still winter. The weather could not have been worse; snow, rain and strong winds reduced visibility almost to nothing. In the month of March, 1902 he finally found the most suitable geographic position to install a new radio station on the island of Cape Breton at Table Head a locality situated a mile away from Glace Bay which received great economic benefits from this, changing it from an agricultural town to an important commercial centre.

The radio station at Table Head became larger and three times more powerful than the one at Poldhu. All the radio stations were built on the Atlantic coasts and were thus always exposed to the danger of storms and winter gales. The unfavourable climate was one of the greatest problems for the radio stations. Guglielmo regularly sailed across the ocean to check that everything worked and to improve the transmissions; the yacht's radio station became his own personal laboratory. In this way he maintained the long distance contact between the transmitting and receiving stations, built on two promontories on either side of the Atlantic.

Guglielmo told me that his life at that time was particularly busy; he personally supervised the building of the new radio stations, just as he chose all his assistants himself. He was full of optimism and enthusiasm and he did everything he could to inspire the same feelings in the people who helped him. Mr. Vyvyen worked for a long time in Canada; he was in charge of the radio stations during Guglielmo's absence. In an article he wrote about my husband, he said: "Only those who have worked with Marconi throughout these four years realize the wonderful courage he showed under frequent disappointments, the extraordinary fertility of his mind in inventing new methods to displace others found faulty, and his willingness to work, often for sixteen hours at a time, when any interesting development was being tested." One of Guglielmo's characteristics was his ability to choose the right person for each job. He and Vyvyan worked together hour after hour at the Poldhu, Clifden and Table Head radio stations. Guglielmo was known as an approachable person who had faith in those who worked for him. He thought sincerity was very important, but he was also sensitive enough to know when it was better to keep quiet.

Here, by the way, I should like to mention that Guglielmo spent the summer of 1902 on board the ship, the *Carlo Alberto*. Sailing between Russia and North Africa, he carried out tests on the "magnetic detector", a technological jewel he had invented which was unaffected by the ship's movement but so sensitive that it could pick up even the faintest electric waves. In October of the same year he was once again on board the *Carlo Alberto,* which was to take him from England to Cape Breton in Canada. He perfected his magnetic detector while he was in constant contact with Poldhu. When he arrived in the town of Glace Bay he was met by a myriad of boats and hundreds of people who had come to welcome him, including his loyal assistants, Mr Kemp, Mr Paget and Mr Vyvyan. The *Carlo Alberto* continued its voyage and anchored in the port of Sydney; here, too, Guglielmo was received with great enthusiasm and gratitude by the local press and by the members of the Sebastian Cabot Society, an important Canadian association.

Canada still feels a debt of gratitude to Guglielmo because his invention gave work to the inhabitants of the island of Cape Breton both to guarantee the constant functioning of his radio stations and to construct and perfect the equipment he invented. For his part, my husband was grateful to the Canadian government and the people of Cape Breton. When he spoke to me about those years, Guglielmo made me feel the great emotion he had felt every time he crossed the Atlantic and admired the natural beauties of the Canadian coasts when they were lit up by the summer sun.

In January, 1903 Marconi arrived on Cape Cod where the South Wellfleet station had been rebuilt with a new set of towers supporting a "V" shaped aerial, modelled on the ones at Poldhu and Table Head. He established a radio link to send messages from the Cape Cod station to Poldhu via Table Head and viceversa. In spite of the bad winter weather, on 19th January 1903 Guglielmo successfully transmitted a message from President Roosevelt to King Edward VII. The signal from Poldhu acknowledging reception of the message did not come back to Cape Cod via Table Head as expected but directly from Poldhu to Cape Cod.

The President's message read:

His Majesty, Edward VII
London, Eng.

In taking advantage of the wonderful triumph of scientific research and ingenuity which has been achieved in perfecting a system of wireless telegraphy, I extend on behalf of the American people most cordial greetings and good wishes to you and all the people of the British Empire.

Theodore Roosevelt

Wellfleet, Mass., Jan.19, 1903.

The reply from the King came back:

Sandringham, Jan. 19, 1903
The President,
White House, Washington, America

I thank you most sincerely for the kind message which I have just received from you, through Marconi's trans-Atlantic wireless telegraphy. I sincerely reciprocate in the name of the people of the British Empire the cordial greetings and friendly sentiment expressed by you on behalf of the American Nation and I heartily wish you and your country every possible prosperity.

Edward R. and I.

My husband enjoyed telling me about that time when he was young, when the new radio service on ships also began. Cunard was the first shipping company which had confidence in him and agreed to his proposal to install it on their transatlantic liners. This was a success for Guglielmo as a businessman and profitable for the Marconi Company.

In 1905, six miles from Table Head and a little further inland, he set up another radio station which was given the name of "Marconi Towers". The spectacular sight of this larger and more powerful radio station could be admired from at least fifty miles away. In the same year Guglielmo also began work on a new radio station at Clifden in Ireland to communicate with Cape Breton. The site he chose was on a plain near the beautiful Atlantic coast of Connemara in south-west Ireland. It was well-equipped and more powerful than the one at Poldhu, although less so than the station at Coltano, the most important one, which was in Italy near Livorno. The Clifden station was inaugurated on 15th October 1907 when a successful radio transmission with Glace Bay was carried out. Guglielmo was particularly pleased by the success of the Clifden station since Ireland was his mother's native land.

In February, 1908, Guglielmo started a permanent commercial radio service. From that moment he began an activity that he knew he could extend all over the world. He told me that the following year a fire broke out at the Glace Bay radio station caused by the excessively high tension. Fortunately, there were no victims. Guglielmo stayed there for many months, supervising the reconstruction personally, making it safer and installing more efficient and powerful machinery. In 1913, twelve radio-telegraph operators were sent to the new radio station at Louisbourg, built by Guglielmo once again on the island of Cape Breton. In 1919 in a radio broadcast from Ireland, the human voice was heard for the first time at Louisbourg.

At the beginning of the First World War, Guglielmo was in Canada; he returned to Italy to put himself at the disposal of the army as a volunteer, offering his help to improve the radio transmissions. He was given the rank of captain. Guglielmo often spoke to me of the time in 1919, immediately after the end of the war, when he was sent by the Italian government to the Versailles Peace Conference as a member of the Italian delegation. The Conference had the task of drawing up the official peace treaty between the principal victorious nations (Great Britain, France, Italy, USA and Japan) and the defeated ones (above all, Germany, which was to suffer extremely harsh terms) in the great war which had just ended.

The Italian delegation was led by the Prime Minister, Vittorio Emanuele Orlando and by the Foreign Minister, Sidney Sonnino (who at a certain moment abandoned the Conference in protest against the terms which were far from favourable for Italy). On 28th June, 1919 the Italian delegation, together with the others, signed the final peace treaty and my husband's signature can still be recognized today together with the others on the final document which is kept in the archives of the French State. He was very proud of this fact although privately he used to say with disapproval that it had been a "mutilated" victory for Italy.

# FOWEY HARBOUR

We went to Cornwall every year; we often visited Torquay, Falmouth, Plymouth and Dartmouth, the home of the Naval Academy for Officers of the Royal Navy which Guglielmo admired so much. All the same, he was proud to be a Rear Admiral of the Italian Navy and thought highly of our own Naval Academy at Livorno. When we were at Dartmouth we often entertained high ranking naval officers who enjoyed coming on board the *Elettra.*

However, our favourite place, the one we loved the most, was Fowey Harbour. Fowey is a fishing village on the south-west coast of Cornwall, built on both banks of the River Fowey near the estuary. The sea and the river at Fowey are green and so are both banks of the river. In summer the leafy willows droop gracefully into the water.

The village is very old and the houses are small and low, lined up along the river bank. Every family has a fishing boat moored in front of the house and in order to get into their houses the fishermen climb up a wooden ladder attached to the wall and held up by ropes. The only inhabitants of this quiet and peaceful village are fishermen and their families whose hard-working lives revolve around fishing. They have been living there happily for generations. The old church and the little flower-filled cemetery lie not far away. This is the real, old England.

Fowey was a haven for both of us. Guglielmo went on with his experiments while I read a lot. We used to interrupt the peace of those days with short boat trips along the river in the Thornicroft motorboat. There were delightful little inlets where the trees and greenery on the banks were reflected in the still water giving a refreshing and restful feeling. How often during the rest of the year would Guglielmo remember nostalgically Fowey's beauty and tranquillity!

There was only one other person who loved the village as much as we did. This was the great actor, Sir Gerald Du Maurier who had chosen it as his retreat. He had built himself a house on the river-bank which appeared to be in a grotto. The sitting-room was almost at water level on the natural rocks. The rooms were like those of the local fishermen but every comfort was provided. Sir Gerald Du Maurier used to spend the weekends there with his family. It was always a pleasure for us to visit them. Guglielmo enjoyed himself in that typically sea-side house and we spent many happy hours in its informal atmosphere. The Du Mauriers were a charming family and when we were at Fowey they often came to lunch with us on board the *Elettra.* We were all very cheerful. Daphne, Sir Gerald and Lady Du Maurier's daughter, was then a fair-haired and very lively young girl; she used to have fun climbing on the yacht's rigging. She was so graceful and agile that it was a pleasure to watch her. In those days I would never have thought that she would become a great writer and win such popularity with her famous novel “Rebecca”.

# THE ILLUMINATION OF SYDNEY 26th March, 1930

My husband was gifted with a self control and a presence of mind which never deserted him, even on days when important events were taking place.

Guglielmo and I were very close. We complemented one another and for this reason we always wanted to be together. Our mutual love was so strong and profound that we could not bear to be separated for long. During the spring that preceded the birth of our daughter Elettra, my physical endurance was put to the test. When it was certain that I was expecting a baby, Guglielmo thought it was too dangerous for me to go to sea. He did not want to leave me alone and in a serious and firm tone of voice he said to me, "I have decided to give up my research on board the *Elettra.* I will stay with you somewhere on land". I knew, however, that he was working on some important experiments at that time and in order not to interrupt his great work I insisted on taking the risk of going with him. So in March, as usual, I went on board the *Elettra* with him. Since I was expecting a baby I felt terribly sick whenever the sea was rough and I also felt tired and unwell because of the strong vibrations and the noise when the generators which were necessary for my husband's experiments started up. I often had to lie down on the deck. Guglielmo was anxious about me and decided I should have my mother with me. As usual her presence was invaluable and gave us great peace of mind.

Finally, on 26th March 1930 in Genova, my sacrifice was rewarded. I was able to be present at the wonderful, unforgettable moment when Guglielmo, from the radio station on board the *Elettra*, touched the key which simultaneously lit up the thousands of electric light bulbs of the Town Hall and the World

Exhibition at Sydney in Australia. We were on board the *Elettra*, anchored in the little 'Duca degli Abruzzi' harbour in front of the Italian Yacht Club of which we were members.

The evening before Guglielmo went to bed as usual around half past ten and fell asleep at once, just as if he had nothing special to do the next day. He slept soundly without ever waking, breathing deeply and regularly, until the morning. He woke up at half past seven, well-rested and calm. He had his usual English breakfast, consisting of tea with milk, two eggs boiled for precisely three minutes (they had to be just so) and toast with butter and marmalade. Then he went to the yacht's radio station, his laboratory; looked at the barometer, checked the temperature and concentrated on the short-wave experiment which was planned.

The appointment with Sydney was fixed for eleven o'clock in the morning which was eight o'clock in the evening in Australia. I can still see the *Elettra's* big radio cabin with the tall heavy apparatus of the short-wave "beam system" which formed the receiving and transmitting station--the only one in existence in the world at that time--as well as the large high-kilowatt valves; on one side of the cabin there was a table with a push-button, radio sets and headphones. Guglielmo was standing by his desk, calm, smiling and sure of himself; he had complete faith in the successful outcome of his research which, by then, he had been working on for a long time. My husband's experimental scientific work and the preparations to set up this exceptional contact with Australia had been long and tiring. Various dignitaries were present, including the Prefect and the Mayor of Genova, the British and Australian Consuls, as well as the Captain of the *Elettra*, Girolamo Stagnaro, the ship's officers, the Chief Engineer, Giuseppe Vigo, as well as many representatives of the Italian and foreign press. Everyone was anxiously awaiting the outcome of Marconi's exceptional experiment which had kept the whole world in suspense for the past few days. It was destined to mark a very important stage in the field of radiocommunications.

Guglielmo sent a message by radio telephony to Australia, where the excitement about the event was intense. Speaking in English, he said, "It is a great pleasure for me to perform the ceremony of officially illuminating the symbol of the new project of the Association for the Radioelectric Development of New South Wales. The switches of the lighting system of the Town Hall of the City of Sydney will be activated by means of radiotelegraphy from the yacht, *Elettra*, which is at present in the Mediterranean at Genova. By pressing a key on board the *Elettra* I will automatically release a "wave train" through the "beam system" of the radio station in England, which will be received practically instantaneously at Rockbank in the State of Victoria in Australia. This impulse will be automatically re-transmitted on Australian territory over five hundred and fifty miles in a straight line to the Town Hall of Sydney, where I will provoke the intake of energy in the light circuit".

I was standing not far from my husband. In the meantime, the group of journalists on the deck of the *Elettra* had become a crowd. At eleven o'clock sharp Marconi pressed a key, and immediately in the silence of the radio station, through the headphones, we heard the voice of Mr Fisk in Sydney, "All well! Wonderful! The leaves are falling here and there in Italy it is spring". Guglielmo turned to me at once, looking deep into my eyes as he always did when he wanted to show me his love; then he smiled at me happily and embraced me. He was grateful to me for the long time I had spent at his side during the preparations for the experiment; all the more because I was expecting Elettra at the time.

From the receiving station in Sydney they informed us that the illumination had been a complete success. Thousands of light bulbs had lit up instantly, to be greeted by the enthusiastic applause of the crowds of people inside and outside the Town Hall. Those present on board the yacht Elettra were delighted. The Australian Consul presented me with a big bunch of red roses as a tribute from the women of his country. I was very touched. The event made headline news and there were long articles in the press all over the world praising the works and the genius of Guglielmo Marconi.

Today, witnessing the continuous progress of science and living in the era of space flights, I can understand more and more clearly what Guglielmo felt in his heart that day. I knew he was proud to be the only person who could communicate directly from Genova with Sydney in Australia in the antipodes and obviously also with other places that were closer. All this from his laboratory, his special experimental radio station, by means of the short-wave "beam system" that he had invented on board the *Elettra*. During his trials Guglielmo said to me: "Just think! Sydney is at almost the furthest point in the world from Genova. That is why I have chosen it." The enormous distance between the two cities did not daunt him. He knew what he was capable of achieving and he was certain that his studies would prove him right. It was really thanks to this conviction of his that he succeeded in what was a unique experiment in those days; something other scientists and researchers had never even dreamed of. A new era had begun for humanity thanks to the type of contact that had just been made with Australia.

Apart from his confidence in his own work, another characteristic of Guglielmo's genius was his ability to organise everything meticulously, without forgetting a single detail. He used to say to me: "The smallest details are indispensable for a perfect result". He realised that he was privileged and exceptional but at the same time he faced the important events of life calmly and sensibly. He knew perfectly well that his work was unique and his inventions all his own but in spite of this, with the humility of the truly great, he often said: "This is a gift that has been given to me by God!". Since I understood his deepest sentiments so well, I knew how happy and satisfied he was at having succeeded with the illumination of the Town Hall of Sydney in creating another important invention for the benefit of humanity.

Talking about the *Elettra*, Guglielmo always used to say to me: "This is a yacht for work, not just for pleasure". We went on adventurous voyages together, defying the terrible Atlantic storms. I remember that Guglielmo and I crossed the Bay of Biscay on board the *Elettra* seven times and the sea was almost always rough as is so often the case in those parts. Only once did we find a calm sea. At the end of these voyages I would arrive in port tired and thin and the same was true for the members of the crew. The only person who showed great physical endurance was my husband. More than once during our voyages Guglielmo told me that our love was a source of strength for him and that it also gave him the inspiration for new ideas. This encouraged me to go with him wherever he went.

When we were in Genova, while he was preparing the experiment of the illumination of Sydney, we were often invited by our Genoese friends to their beautiful palaces. I remember a wonderful evening at the Marchese Raggio's castle where we were guests of the Marchesa Thea, a Spinola by birth, whose second husband was the Marchese Giuseppe Cattaneo della Volta. We had also become friends of the Marchese Marcello Gropallo and his wife Rosa, who often invited us to their famous villa "Lo Zerbino", with the Marchese Rodolfo Pallavicino and his wife Maria, the Marchese and Marchesa Vistarino and the Marchese Carlo Raffaele Bombrini. We often went to the Marchese Franco Spinola's splendid villa by the sea near Rapallo which was called "La Pagana". He always received us wearing a naval officer's uniform. As a Dame of Honour and Devotion of the Sovereign Military Order of Malta (S.M.O.M.) it is a pleasure for me to remember that this villa which we used to visit was later donated by the Marquis Spinola to the S.M.O.M. We ourselves enjoyed inviting our friends to lunch or dinner on board the *Elettra*. I remember that among them was a young naval officer, Luigi Durand de La Penne, the future admiral who was awarded a Gold Medal for gallantry for his exploits during the Second World War.

After the illumination of Sydney, having stayed at Genova for about two months, we sailed along the Maremma and the Roman coast as far as Fiumicino, off Mount Cavo in the Roman hills. Suddenly a storm blew up, just as Guglielmo was carrying out an important experiment. He immediately gave orders to Captain Stagnaro to take shelter in the port of Civitavecchia, which was the closest. The pilot of that area, knowing that I was on board, was anxious because of my condition and immediately came to meet us. He was the father of ten children! Guglielmo and my mother were very worried about me but all went well. My husband was so upset by the risk I had run that he decided to disembark. By then I was seven months pregnant with Elettra.

1. First signed photograph that Marconi gave to his future wife, Maria Cristina Bezzi Scali.

2. Annie Marconi Jameson with her sons Alfonso, and, Guglielmo to her left--Bologna, 1878.

3. Guglielmo Marconi with his invention--Wireless Telegraphy--1895.

4. Guglielmo Marconi and Maria Cristina, on their wedding day--15 June 1927.

5. Marconi with Maria Cristina leaving the Capitol of Rome, on the day of their civil matrimony, a few days before their religious matrimony.

Marconi in his makeshift lab at St John's, Newfoundland, where he received the first transatlantic wireless signals from Cornwall, 12 December 1901. *Marconi.*

6. (Caption part of the photo).

7. Marconi in his Radio Station, transmitting across the Atlantic. The antennas were called the *Marconi Towers*, located in Glace Bay, Nova Scotia, 17 October 1907. The inventor, Marconi, is seated to the left, and the operator, James Holmes, is receiving a message from the Clifden Ration Station in Ireland.

8. Autographed photo by Marconi the day he turned on the lights in Sydney's Town Hall, Australia, from aboard the *Elettra*--26 March 1913.

9. Marconi in his radio cabin aboard the *Elettra* at the moment he is throwing the switch to light Sydney's Town Hall in Australia--26 March 1930.

10. Aboard the *Elettra*, anchored in the small harbor at Duca degli Abruzzi, Genova, Italy, Marconi throws the switch that lights the Town Hall of Sydney in Australia, with his signature.

11. The yacht, *Elettra*, in the Duca D'Abruzzi harbour in Genova, Italy, with all flags flying in honour of Marconi's having turned on the lights of Sydney's Town Hall, from his radio cabin--26 March 1930.

12. Pontifical mission of 1911 in occasion of the crowning of His Majesty, King George V. Cardinal Gennaro Granito Pignatelli of Belmonte, Monsignor Eugenio Pacelli, Count Francesco Bezzi Scali, Lord Crichton Stewart, and Count Medollago Albani.

13. Marconi with his Eminence, Cardinal Secretary of State, Eugenio Pacelli, at the Christening of Marconi's daughter, Elettra.

14. Countess Bezzi Scali with her son, Antonio, and her little daughter, Cristina--Rome, 1901.

15. Countess Anna Bezzi Scali, born Marchesa Sacchetti.

16. Count Francesco Bezzi Scali, in uniform, was Brigadeer General of the Noble Guards of the Pope.

17. New York, Columbia University, 1927: Chancellor Nicholas Murray Butler confers Honorary Doctorate onto Guglielmo Marconi.

18. Marconi with His Royal Highness, the Duke of Abruzzi-Savoia. Marconi is in full regalia and serving as President for the Royal Academy of Italy--29 November 1930.

19. Marconi dressed as a yachtsman--1928.

20. Marconi and Maria Cristina aboard the *Elettra*--1929.

21. Marconi aboard the *Elettra*--1929.

22. Marconi's new-born daughter, Elettra, in her mother's arms. Looking on is the grandmother, Contessa Anna Bezzi Scali--July, 1930.

23. Studio picture of Marconi with his wife, Maria Cristina, and daughter, Elettra--Rome 1931.

24. Full view of Marconi's *Elettra* at Cowes on the Isle of Wight.

25. Aboard the *Elettra*, 1931: Marchesa Maria Cristina, the nurse holding little Elettra, and Marconi.

26. Marconi with the Duke of Kent, exiting from the Marconi's Wireless Telegraph, Co.

27. Marconi with Pope Pius XI and with Cardinal Eugenio Pacelli at the inauguration of Radio Vatican--1931.

28. Marconi, with headphones, sits in front of the microphone, giving one of his many addresses. The verses beneath the photo is as follows:

    His image will continue
    Among the people of the sea.
    In silence, they bow their heads
    and say with full emotion,
    "He is Marconi."

29. Photo of Marconi dedicated to his wife Maria Cristina: *To my always beloved Cristina, forever yours. Guglielmo--Rome 28 April 1932.*

30. Signed photograph by Marconi during a conference at the Royal Academy of Italy.

31. Marchesa Maria Cristina Marconi wearing the emerald necklace given to her by Marconi at the birth of their daughter, Elettra--Venice.

# THE BIRTH OF ELETTRA

We went from Civitavecchia to Santa Marinella where my cousins the Marchese and Marchesa Sacchetti had their summer villa. It was really pleasant to see our relatives again. Prince Odescalchi used to spend the summer in his castle at Palo. When he heard from my cousins that Guglielmo and I were planning to stay in the area until the birth of the baby he offered us his lovely villa by the sea. We were delighted to accept and rented it for the three summer months. In this way my husband could carry on with his experiments on board the *Elettra* which was anchored out at sea. I have wonderful memories of those days. Villa Odescalchi was built on the rocks outside the town towards Santa Marinella and Guglielmo and I had a magnificent room overlooking the sea. It was here that our beloved Elettra was born.

On 20th July, 1930, the day of our daughter's birth, the waves were breaking violently against the rocks and making a deafening noise. The din did not bother me, in fact I liked it. Guglielmo was with me, encouraging me with his smile. We talked lovingly about the little baby that was about to come into the world and wondered what it would be like. We did not yet know that our child would love the sea, which was so close and made itself heard so violently that day, as much as Guglielmo and I did. I had a beautiful baby girl with blue eyes and fair hair. Guglielmo wanted to call her Elettra like his yacht. We were filled with joy. We had a daughter who made our happiness complete.

Our daughter was christened Maria in devotion to the Virgin Mary; Elettra after the yacht; Elena in honour of her godmother the Queen of Italy and Anna after her two grandmothers, Guglielmo's mother and my own.

Elettra was baptized on 30th July at Villa Odescalchi in a room that was turned into a chapel for the occasion. His Eminence Cardinal Eugenio Pacelli, who was then the Secretary of State came expressly from the Vatican City. He had wanted to bless my marriage to Guglielmo in 1927 but unfortunately at that time he had not been able to leave Berlin where he was the Papal Nunzio. He did us the great honour of baptizing our daughter. The future Pope Pius XII was a very old friend of my family and ever since I was a little girl I had been fortunate enough to meet him often at our house. When he arrived at the villa everyone was struck by his tall figure dressed in the purple cardinal's robes. He was good looking and impressive and he emanated a great spirituality.

Elettra's godmother was Queen Elena who gave her a beautiful diamond medal and a little gold and pearl necklace. Her majesty was represented by a lady in waiting, the Duchess of Laurenzana. The christening ceremony was followed by a reception in the garden. The owners of the villa, Prince and Princess Odescalchi were present. Enzo and Vittoria were close friends of ours and I remember them with great affection. Our friendship continues with their children. Naturally all my closest relatives were there: Prince Orsini, Prince and Princess Barberini, Marchese and Marchesa Sacchetti, Marchese and Marchesa Serlupi Crescenzi, my aunt Maria, my father's sister who was married to Marchese Guglielmi D'Antognolla. Among our friends were the Princes and Princesses Colonna, Massimo, Chigi, Borghese, del Drago, Ruspoli, Lancellotti, and Torlonia, Duke and Duchess Lante delle Rovere, Duke and Duchess Sforza Cesarini, Marchese and Marchesa Theodoli and the Governor of Rome Prince Francesco Boncompagni Ludovisi with his wife Princess Nicoletta.

The weather was fine that day and the sea was calm. After the christening Guglielmo asked Cardinal Pacelli if he would like to hear some messages from Australia in the radio station on board the *Elettra*. This was still a great novelty as the first official communication had taken place only a short time before. The Cardinal, who was always very interested in my husband's experiments, was delighted to accept the invitation. The *Elettra* which was anchored in front of Villa Odescalchi had all her flags flying and looked more beautiful than ever that day. His Eminence, accompanied by Guglielmo, went out to the yacht in our motorboat and was ceremoniously welcomed on board the *Elettra* by the Captain and the crew. He stayed for some time in the radio station and listened with interest to Guglielmo's explanations and to the radio signals and conversations from far away countries.

Just a few members of my family and our closest friends stayed on after Cardinal Pacelli's departure. Guglielmo took Prince Odescalchi's children, Livio, Anna, Ladislao and Alessandro on board the *Elettra* so that they could visit the yacht and listen to the radio messages that came from far away. Guido and Maria were too little to go. I can still see my husband's happy expression at the end of that exceptional day.

Keeping a promise he made us at the time, and therefore as an exception, a few years later Cardinal Pacelli who by then had become Pope Pius XII administered First Communion and Confirmation to our daughter Elettra in the Matilde Chapel in the Vatican. When his Holiness Pius XII died in 1958 it left a great void in our lives.

# VILLA GRIFFONE

Not long after our marriage I asked my husband "What made you think of inventing radio-telegraphy?" He answered that ever since he was eight years old he had been sure that he would invent something very useful for humanity which would make him famous so that he would be considered different from other men. He did indeed succeed in doing so. With his genius and perseverance he was able to abolish distances, accomplishing the miracle of instant communication through space. He told me that when he was fourteen years old, using some wires and a broom-stick, he built the first antenna! He established the first contact between the air and the earth from the stone balcony at the front of the house where he spent his holidays with his parents on the sea front of Livorno.

Guglielmo had heard about Hertz's discoveries in the field of electro-magnetic waves; he was very excited by these discoveries and immediately had the idea of using them in his experiments on wireless communications over a distance. He understood the importance of the herzian waves straight away but having a deeper and more complete personal intuition about this phenomenon he decided to make a thorough study of them in order to develop and extend his research. Although he was so young, he was very determined and even during the summer months while he was enjoying his holidays in the country or at the seaside he always dedicated hours and hours to his scientific research work.

During the winter when the weather was colder the Marconi family used to move for a month or two to the milder climate of Livorno or Florence. In Livorno they would rent a flat in Viale Margherita with a view of the sea. Here Guglielmo met Professor Vincenzo Rosa, a professor of physics, who immediately noted that the young Marconi had an aptitude and a passion for physics and saw that his studies were very individual and quite different from those of his contemporaries. He realized that here was a brilliant mind whose depth and breadth of spirit and scientific ability he admired. They also used to meet sometimes in Bologna. In Livorno, in the State Technical High School at No. 9, Via Cairoli, there is a memorial plaque in honour of Guglielmo Marconi to commemorate his attendance at the school and his frequent visits there. Another memorial plaque was dedicated to him by the Scientific High School of Florence in Piazza Santa Trinità. Both cities wanted to commemorate his studies and residence there. The first time I returned to Italy soon after our marriage I went with my husband to see these two plaques. I remember so well his saying to me with a smile, "It's a rare thing for a person to read his own memorial plaque during his lifetime!"

The following are some of the events of his youth and his first inventions. My husband came from a distinguished and respected family from Bologna. His father Giuseppe Marconi was a well-known landowner with extensive agricultural estates both in Emilia and Romagna. A successful businessman, he took a particular interest in all his farm-workers because he wanted them to live in respectable comfort with their families on his land. Guglielmo's mother Annie was born Annie Jameson, a member of the Jameson family of Irish whiskey distillers. Her grandfather John Jameson was born in Alloa in Scotland and became Sheriff Clerk of Clackmannonshire. He went to Ireland and bought an interest in a distillery in Bow Street in Dublin. Her father Andrew set up a whiskey distillery in Fairfield near Enniscorthy, County Wexford in the south-east of Ireland.

Andrew and his family lived in a beautiful old moated house, Daphne Castle, surrounded by a park near Enniscorthy. Here Annie was born in 1839, one of six children. She was a lovely girl, charming and vivacious. The whole family had a passion for music. It was their favourite occupation and they used to play different instruments together in the evenings. Annie played the piano and she also had a very beautiful soprano voice. To her family's disapproval (it was quite out of the question for a young girl of good family to become an opera singer in those days) she was offered an engagement to sing at the Covent Garden Opera House in London. She was forbidden to accept but after much discussion and argument she was allowed to go to Italy to study singing with a famous teacher of the time. The Jameson whiskey firm had business contacts with a Bologna banker called De Renoli and Annie was sent to stay with his family. De Renoli's daughter had died while giving birth to a son, Luigi, and their widowed son-in-law Giuseppe Marconi spent much of his time with them. He was charming and lively with a good sense of humour and he and Annie soon fell in love.

When Annie returned home to Ireland she asked permission to marry Giuseppe. Her family was just as shocked at Annie's choice of husband as they had been at the thought of her singing at Covent Garden and they refused their consent to the marriage. Annie was ordered to forget him and, apparently obedient, she remained at home and led a social life going to parties and meeting suitable young men approved of by her family. However, she continued to correspond secretly with Giuseppe. She had a far stronger will than anyone realized and her mind was already made up. When she came of age she ran away from home and crossed the stormy waters of the Channel to France while Giuseppe drove across the Alps in his carriage. They met in Boulogne-sur-mer, a romantic town by the sand dunes and were married there on 16th April 1864. She and her family were reconciled after the birth of her first child, Alfonso the following year. Nine years were to go by before the birth of her second son, Guglielmo.

Annie transmitted her love of music to her sons. The two brothers were both gifted musicians. Alfonso played the violin while Guglielmo was a talented pianist. His mother taught him to play from an early age and she would sing while he played. All his life whenever it was possible he arranged to have a piano nearby and if he had a spare moment he would sit down at the piano and his beautiful hands would fly over the keys. He could read music at first sight and often improvised or played music he knew by heart. He once said: "I love music. I had a serious musical education from my mother. Playing the piano and developing my sensitivity for harmonious and delicate notes has been of great help to me in the scientific field".

Annie lived happily in the beautiful ancient town of Bologna with her two towers, medieval palaces and porticoes and characteristic pink stone that lights up in the sunset. When the Marconi family was not in Bologna they used to spend part of the year at Villa Griffone, Giuseppe's country house. Villa Griffone is a Seventeenth Century villa in a magnificent position on top of a hill which catches the breeze even in summer, in the district of Pontecchio, a village in the district of Sasso Marconi near Bologna. Over the centuries it has undergone various transformations; it is a bright comfortable house with many windows opening onto a pleasant landscape surrounded by fertile fields and vineyards and shaded by lemon and chestnut trees. The walls are very thick, the rooms are large and there is a spacious hall, a beautiful stone staircase and the floors too are flagged in stone.

When he was at Villa Griffone Guglielmo spent many hours of the day shut up in the granary which he had turned into a laboratory, studying and making experiments, surrounded by his rudimentary instruments, the same ones that can still be admired today, some in London in the Science Museum in South Kensington and Chelmsford, others in Milan in the Leonardo da Vinci Museum, and some in Rome at the Ministry of Posts and Telecommunications. All this meant that he could not always enjoy the company of his mother, father and older brother Alfonso; he used to shut himself away, concentrating on the work that was to bring his passionate research to a conclusion. He was so busy that for two years the family had to give up their usual winter move to Livorno and Florence.

At that time many physicists and researchers in different parts of the world were trying to construct instruments which would generate and detect the radiation of electro-magnetic waves, study their properties and pick them up at a certain distance, starting from the theory of electric waves developed by the Scottish researcher Maxwell: among these, Righi in Italy, Edison in America, Hertz in Germany, Lodge in England and Branly in France.

Guglielmo knew about the studies of Professor Augusto Righi, an expert in physics who was carrying out experiments in a laboratory in the hills on the other side of the River Reno from Villa Griffone. He decided to go and see him to discover whether Righi had acquired any useful information for the research which interested him so much. My husband told me that he went on horseback, since the climb was so steep, but that was the first and last time he went because although he felt respect for the scientist he saw that Righi was very skeptical about his research. Guglielmo realized at once that the Professor's work was quite different from what he himself had in mind and was, in fact, already achieving.

He told me he decided to turn to new systems of research which the above-mentioned scientists had not thought of. With the profound intuition which was a characteristic of his genius he was convinced that another type of electromagnetic waves, different from those used up to then in the laboratories, would be necessary to receive communications from far-off stations and surmount the natural obstacles on the way. This intuition made it possible for him to succeed in trasmitting signals over various distances. So, strengthened by his belief in his own ideas, he went on with his work alone, never consulting anyone else and making totally unique experiments at Villa Griffone, absorbed in his research to find the way to use these waves. At last Guglielmo achieved his purpose: to communicate freely in the void of the ether without the use of wires. He reached his goal where the other scientists still searched and searched...

One of my husband's first important inventions was his new improved "coherer". Apparatus consisting of a detector of high-frequency discharges using a glass tube and a conducting powder had already existed since 1884. This apparatus, however, although it had been modified and perfected by various scientists for other purposes was in Guglielmo's opinion very primitive and ineffective. He wanted to apply this instrument to long-distance wireless transmissions. After much research and many ingenious modifications, Marconi transformed the apparatus, reducing the diameter of the tube, inserting two silver electrodes very close together and filling it with a powder consisting of nickel and silver filings with traces of mercury, all in a vacuum. Guglielmo explained these things to me simply and clearly. These modifications so greatly increased the effectiveness of the apparatus that he himself considered it to be a new instrument of his own invention. Thus perfected, the "coherer" acquired an unequalled sensitivity and resistance. This is the reason why he had no hesitation in calling it, as I have already said, one of his first important inventions, a true creation of his own, thanks to which the capacity of transmissions from one station to another was gradually increased. The signals could be picked up not only indoors but also by a receiver system placed outside the house. I think it is important to tell all this because Guglielmo spoke to me about it more than once. He wanted me to know the truth. He trusted me and he knew that I in my turn would speak about it to the future generations.

My husband always remembered with pride his first success when he was still an adolescent. He told me that one day at Villa Griffone, after much experimentation, he called his mother. "Come with me", he said. "I have something to show you. A surprise." Annie Jameson Marconi was blonde with beautiful long hair and soft blue eyes. An English cousin of hers told me that she was an intelligent and determined woman. She had very high principles and her whole life was dedicated to bringing up her two sons Alfonso and Guglielmo who was nine years younger than his brother. She was the first to realize that there was something special about her younger son, a sensitive and thoughtful child, quite different from his older brother in character, intellect and tastes. Mother and son understood each other and got on very well together. He told me, "I knew she would never doubt me and would believe in my intuition. I wanted to show her that by just pressing the button of a bell, fixed to the floor in the centre of the room, it was possible to ring another bell in the next-door room with the door closed. Annie Marconi watched this experiment and was astonished. It was almost a game but she took it seriously. She called her husband so that he too could see this demonstration. Giuseppe Marconi watched very carefully and shook his head in perplexity. Then he began to search all around, thinking that there must be a hidden electric wire. He looked carefully along the walls of the two rooms and investigated the floor, lifting up the carpet. He could not understand how the bell could ring in the next-door room without a connecting wire and with the door closed. Giuglielmo found all this incredulity very funny.

He told me that his father was rather skeptical about his experiments but he did not feel offended or allow himself to be influenced by this because he was absolutely certain he would succeed. He respected his father but went on with his studies with his mother's support. She even spent two winters in Villa Griffone at Pontecchio so as not to leave her son alone during those years of hard work. Some old people from Bologna still remember it. Although the fires were lit in the fireplaces, the intense cold in the villa was hard to bear. Although his mother felt the cold she wanted to make this sacrifice for Guglielmo so that he could develop whatever he had in mind. Alfonso, his older brother, told me that when Guglielmo was in the granary he was so taken up with his studies and experiments, so determined to succeed, that he even forgot to eat and became very thin. One can see this clearly in photographs taken at the time. He shut himself away for days on end until late in the evening. He did not have meals with the family and sometimes his mother, worried about him, would bring him a bowl of nourishing broth which he would refuse. He never touched a mouthful until he had achieved his aim. His father used to grumble about this younger son, this "eccentric" who wasted time and money with his coils and wires, shut up there in the silk worms' room which he had transformed into a laboratory.

All this happened before 1895, before the famous experiment of Pontecchio which connected the granary of Villa Griffone to the country on the other side of the hill beyond the garden in front of the villa along the now historic avenue. When Guglielmo told me about those days, he always told me how deeply grateful he felt to his mother. Once they went together to the Sanctuary of the Madonna on Mount Oropa near Biella. He stood for a long time admiring the view, thinking that God had put the forces of nature at man's disposal and he felt sure that with His help he would succeed in exploiting them for the good of humanity. These are his own words, "The free paths of space for the transmission of human thought have had a great fascination for me ever since that moment. Unlimited sources of inspiration for new achievements for the benefit of humanity exist in them". A smile lighting up his face, my husband told me, "When I told one of our farm-workers Marchi to go with my brother Alfonso to the other side of the hill and listen for the first signal I sent from Villa Griffone I asked him to fire a shot the moment he heard it. I sent the signal of the letter "S" with the transmitter I had built for this experiment. On the other side of the hill I had set up a receiver which I had also built myself that picked up my signal perfectly. Just think, my dear Cristina, what enormous satisfaction I felt when I heard the shot after working so hard to achieve my aim: to communicate in the open, even surmounting obstacles!" In that historic moment, as he himself said, wireless telegraphy was born. I, too, understood that this perfectly successful experiment had marked the beginning of a new era which was to transform our whole existence. Starting from this first positive result, Guglielmo continued to develop and perfect his invention. He succeeded in transmitting over greater and greater distances also by using the power of conduction of the earth which was then unknown. An extraordinary intuition of my husband's in the field of the propagation of electro-magnetic waves concerned the conductivity of the earth in long-distance transmissions. With his brilliant mind he invented and perfected the antenna-earth system, the essential key to the development of the entire radio-telegraphic system.

In the same year, 1895, after his first demonstration experiments at Pontecchio, Guglielmo showed his invention to the military authorities and representatives of the Italian Government in Rome but unfortunately without success. "They didn't believe me!" he told me. There is an eloquent little picture by Bemporad in the "*Domenica del Corriere*" of the time which illustrates this. So he said to his mother: "Mother, you are English; let us go to London". Wasting no time, he collected together the various pieces of his scientific instruments in a big wooden crate which he took with him and he and his mother left for London. Guglielmo had always been very fond of England which he thought of as his second home and which he had often visited when he was a boy. He arrived there full of confidence and enthusiasm with his exceptional invention for wireless communication through the ether. In London, Guglielmo was made welcome and the importance of his invention immediately understood. He was glad to meet his cousin Henry Jameson Davis again. Henry was a good-looking man with a very exuberant character, a few years older than him. Guglielmo was always grateful to him because of the kindness he showed him during this stay in London which was so important for his future. I remember that when I went to London with Guglielmo many years later I too met his cousin Henry Jameson Davis who always showed his sincere affection for my husband. He was Colonel of the Regiment of Irish Volunteers and wore his smart uniform with pride.

Thanks to a letter of introduction from A.A. Campell Swinton, an eminent electrical engineer of the day, Guglielmo was able to show his invention to Mr. William Preece, then Chief Engineer of the British Post Office, who immediately understood its enormous scientific value: it would make it possible for the Great Britain of Queen Victoria to communicate with its Dominions all over the world quickly and in secret. From then on--this was 1896--he lived more in London than in Italy. He recognized that the English had believed in him and recalled with emotion his first experiences in radio transmission: one from the roof of the General Post Office, another from the roof of St.Thomas's Hospital across the Thames and still another from the "Needles", rocks near the Isle of Wight, across the Solent.

Guglielmo told me that after the world-wide recognition of his discovery, his father felt very sorry that he had ever had doubts about his son's success. When the newspapers spoke of him and his inventions, he cut out the articles and the photographs and kept them carefully, regretting that he had not believed in him right from the beginning.

My husband told me some other stories about his youth at Villa Griffone. I remember one in particular; about the bicycle. He wanted one that so he could go to the nearby village, which is now called Sasso Marconi, to get the materials he needed for the construction of his apparatus there in the granary. He had no money of his own to spend and his father was strict and did not approve of buying things which seemed to him to be strange and useless. Finally Guglielmo persuaded him to lend him twenty-five lire and so was able to buy himself the longed-for bicycle. He often used it to go to Sasso and get what he needed immediately. As soon as he could, to his great satisfaction, he repaid his father the twenty-five lire.

Years later I often went to Bologna with my husband and I loved visiting the places where he had spent his summers as a boy with his parents and where he had begun his first experiments in radio-telegraphy. One day in September, 1928 Guglielmo took me to the church of Pontecchio. The church had been decorated especially for our visit; there were flowers everywhere and they had prepared a stool for us to kneel upon, covered in red damask silk, which was usually kept just for weddings. Many people from the village and the neighbourhood were present and when we came out of the church everyone greeted us joyously, clapping and cheering as we went by because Guglielmo, like his father, was much loved by the farm-workers and employees of the Marconi estates.

My husband wanted me to meet the Parish Priest, Don Domenico Calzolari, who had been there ever since Guglielmo was a child. The priest, who was now ninety years old, remembered him very well and told me: "When I went to visit the Marconi family at Villa Griffone for the first time, I took a lamb as a present, with a little bell tied around its neck with a red ribbon. How Guglielmo jumped for joy! He was a beautiful child of about two or three, with fair hair and blue eyes. He often came to Pontecchio with his father in a light four-wheeled gig, drawn by two ponies with smart red English harness, given him by his mother.

Guglielmo had very happy memories of the summer months that he spent with his parents at Pontecchio during his childhood and adolescence. One of his best friends was the little Tettè Malvasìa whose family were friends and neighbours there in the country. He loved the long warm days and said that the summer was never too hot at Villa Griffone. He told me that in the afternoons he used to go for long walks on their land and as he walked he always thought about the possibility of communicating over a distance using electro-magnetic waves. He would throw little stones into the ponds to study the phenomenon of the concentric circles that formed in the water; he always had the propagation of waves in mind. He stuck sticks in the stream to test the antenna which upset the women who came to do their washing there. He often climbed to the top of the hill to admire the sunset. My husband had the soul of a poet and the beauties of nature moved him deeply. A beautiful sunset or the changeable appearance of the sea could send him into raptures. Sometimes as we sailed on board the *Elettra* he held me close as we stood on the bridge and together we watched the stormy sea in silence and admiration.

# THE REGATTAS AT COWES

Every year, during the summer season--usually in the first week of August--we used to go to Cowes for the Regattas.

It was a very important date for Guglielmo and me. The weather was usually bad. The waters of the English Channel were green and cold. Nevertheless, we loved "Cowes Week". When we arrived at the Isle of Wight we would anchor the *Elettra* off Cowes, opposite the Castle of the Royal Yacht Squadron.

I first went to Cowes in 1927, about a month after my wedding to Guglielmo. Aboard our Isotta Fraschini motorlaunch, we drew up to the elegant private landing jetty of the Royal Yacht Squadron, which was considered the most important yacht club in the world.

Guglielmo looked even more English than usual on that occasion. He wore, with sober elegance, the clothes customary for members of this club: white flannel trousers, a double-breasted blue blazer with the emblem of the Royal Yacht Squadron on its gold buttons, a stiff white collar, a blue club tie and a yachtsman's cap bearing the embroidered badge of the Royal Yacht Squadron. In the small town of Cowes there was a tailor who only did work for the members of the Squadron; a detail which may seem superfluous but which Guglielmo considered important because in his opinion it showed the prestige of what was considered the most exclusive yacht club of the age. To be a member was thought a great honour.

It was a very elegant meeting-place where the most important personalities from every part of the world could meet. There were just over fifty members, almost all English, and all of whom had to own a yacht. The President was King George V. There were hardly any foreigners and Guglielmo was the only Italian and proud of it. Conversation there was exclusively about yachts, regattas, the sea and sailing.

The yachts that came to watch the regattas were sometimes as many as five hundred. It was a fine sight to see the the flags of all the different countries flying from their sterns. Many yachts arrived from various European countries and also from the United States; many had steam engines and their owners lived comfortably aboard, just as we did on board the *Elettra*. I remember the *Candia*; the *Phantom*, a schooner belonging to the Guiness family; Lord Astor's *Zephire*, looking like a transantlantic liner; the French *Airbale*; the Duke of Westminster's *Gironde* and the *White Heather*, a beautiful sailing boat.

How Guglielmo and I enjoyed watching the regattas! We were kindly offered a prime anchorage in front of the Castle. From the Elettra we could see the yachts which were taking part in the races passing right in front of us. They were beautiful. As they raced, they leaned to one side with their great white sails swelled by the strong, cold Channel wind. The international regatta, then as now, was called the Fastnet.

I can still see King George V at the helm of his racing yacht, the *Britannia*, dressed in a blue blazer with his yachtsman's cap. I watched him with respect and admiration. But Sir Thomas Lipton's *Shamrock* outdid all the other yachts and always came home first. It advanced majestically, with spread sails, cutting the waters of the Solent with its emerald green bow.

The crews which took part in the races were very skilful. They carried out the manoeuvres with speed and grace. To take advantage of the wind the sailors would lie down almost flat on the deck, holding on tightly so as not to slip into the water when the yacht leaned on its side. It was a hard and fast rule during the regattas that if a sailor fell into the sea the yacht could not stop but had to continue racing. The currents of the Solent were very strong and dangerous. Guglielmo considered it almost impossible to save a man who fell into the sea there. We were thus full of admiration as we watched the actions of those extraordinary seamen.

During Cowes Week, the Royal Yacht was home to King George V and his entourage and was anchored a little further out to sea, isolated from the other yachts. In their day, Queen Victoria, Prince Albert and the Prince of Wales had enjoyed staying aboard the yacht at Cowes. The older members of the Squadron looked at its black hull with affection. Guglielmo himself looked at it with nostalgia because it reminded him of one of his first successes with wireless telegraphy in England. In August, 1898, Queen Victoria, who was in residence at Osborne House in the centre of the Isle of Wight, wanted to use radio signals to keep in constant contact with the Prince of Wales, the future King Edward VII, because he had injured his knee and was confined aboard the Royal Yacht anchored off Cowes. Marconi, the young Italian genius, was sent for as the only one capable of the task and he soon established a reliable radio link, thus satisfying the Queen's wishes. On August 11th Queen Victoria sent the following message: "The Queen hopes the Prince has had a good night and hopes to be on board a little before five". No less than 150 messages were exchanged between Osborne House and the Royal Yacht lying off Cowes over a period of 16 days in August 1898.

In a letter to his father from London, dated 7th September, 1898, Guglielmo wrote: "You may have learned from the newspapers or from what I wrote to Mother that the Queen and the Prince of Wales wanted to see my apparatus and used it for more than two weeks.

I had the pleasure of staying on board the Royal Yacht during the excursions. The Prince of Wales was continually able to telegraph and keep in touch with his Mother the Queen (who was in her palace at Osborne) even when the ship was off-shore, something that would have been impossible without my invention. You may be already aware that the Prince of Wales gave me a very nice and valuable tie-pin as a present. After thanking me very much, he told me he would be very happy to see me any time I wished to visit him".

One day towards the end of Queen Victoria's long and glorious reign, which ended in 1901, the Prince of Wales, later King Edward VII, who was always most kind to Guglielmo, said to him in confidence: "When one has been treated with the greatest consideration for many years, in the end one thinks one is God"; Guglielmo just smiled at this without commenting because he knew the incident being alluded to was when Queen Victoria discovered him in her private garden at Osborne House going about his task; the Queen voiced her displeasure at the intrusion into her privacy; however, when she ordered her staff to "get another electrician" she was told there was nobody else who could do what Mr. Marconi could do; it was only after the wireless messages began to flow that she eventually allowed herself to show some sign of gratitude to Guglielmo, inviting him to lunch with her at Osborne House and she asked to see his radio equipment. The Queen was then 79 and he was just 24; she died just two years later. Guglielmo always remembered both Queen Victoria and the Prince of Wales with great devotion and affection.

When I went to Cowes for the first time all those years later I noticed that the rules of that very traditional society were still in force. Guglielmo wore yachtsman's dress and his club's dinner jacket in the evening. As the wife of a member, during the day I used to wear navy blue with red and white accessories; I wore a blazer with brass buttons with the crest of the Royal Yacht Squadron engraved upon them.

As soon as we arrived in Cowes, Guglielmo took me to Benzie, the shop that specialized in jewellery for yachtsmen and their ladies. He wanted to give me a brooch with the Squadron's red and white flag on it in rubies and diamonds and a gold chain bracelet with little flags in precious stones which formed the name of our yacht in the nautical alphabet. When my daughter Elettra was eighteen I gave her the bracelet as a present in memory of the time when Guglielmo and I sailed together on the *Elettra.*

Towards evening, at the end of the day's regatta we went ashore by motorboat, landing at the Squadron's jetty. It was an extremely smart passage ashore. We were received by naval officers and seamen who were there to greet us with great style and a few appropriate words. Then we went into the Club and it was like going into a sacred place; everyone realized how privileged they were to belong to that exclusive community of sea-loving people to which one was admitted only if in possession of special requisites.

The Squadron's headquarters were in a building called "The Castle" which overlooked the Solent. It was not very large and dated back to the XIV century; inside it was very elegant, very English and very nautical. The members would chat about the regattas of the day and the latest yachts launched and tell anecdotes of adventures at sea, rather as I have heard hunters talking about their shooting exploits.

Guglielmo was very highly thought of in that circle and he too enjoyed being there; he was in good humour, relaxed, smiling and friendly. He liked to remember that it was there he had met the Wright brothers, the pioneers of aviation, who were very charming and famous for their courage. One of them invited Guglielmo to go up in an aeroplane with him. Guglielmo accepted, attracted by the novelty. He told me later that the flight had been exciting but he thought it was still too much of a risk because it was not reliable; I am sure that today he would be delighted to fly in modern aeroplanes.

While we were at Cowes my husband could finally enjoy a real rest. We ladies had tea on the green lawn of the Squadron's garden. In the evening we wore smart clothes appropriate for attending important dinners and receptions. We would disembark from our motorlaunch and go on board the various yachts. Sometimes the evenings ended with magnificent balls at which one could easily meet His Royal Highness, the Prince of Wales, the Duke of Kent and foreign sovereigns and princes who enjoyed the life at sea. There was always a gala evening with the official ball aboard the Royal Navy flag ship anchored in the bay.

Guglielmo and I enjoyed returning our friends' kindness and their invitations, giving dinners and receptions aboard our yacht. Our Italian chef really gave of his best on these occasions. Sovereigns, princes and dukes honoured us with their presence, as well as intimate friends like the Duke and Duchess of Sutherland, Lord and Lady Londonderry, Lord Louis Mountbatten (later the last Viceroy of India) and his beautiful wife Lady Edwina, the Duke of Westminster, Lord and Lady Birkenhead with their daughter Pamela and many others.

Lord Birkenhead usually anchored his yacht, *Mairi*, near ours. He often came aboard the "Elettra" with Lady Birkenhead and their children. He had long conversations with Guglielmo on elevated topics. He was a good looking dark haired man with typically Anglo-Saxon looks. He had been Lord Chancellor and the Minister for the Indies. As a Tory elder statesman he was always listened to with respect when he spoke in the House of Lords. He was a fine orator; he spoke the most perfect English that could be heard in those days. He was very cultured and had a quick mind and a great sense of humour. Guglielmo admired him for these qualities.

When Lord Birkenhead saw me for the first time at Cowes in the summer of 1927, he paid me the nicest compliment I have ever had. Looking at me, he said to Guglielmo: "Marconi, this is the most beautiful discovery of your life".

# PRESENTATION AT COURT

In 1929, I was formally presented to the British Royal Family. One evening we went to Buckingham Palace. Guglielmo was proudly wearing the dress uniform of a captain of the Italian Navy and his many decorations as protocol required on these occasions. One of these decorations was that of a Knight's Grand Cross of the Royal Victorian Order, awarded him by King George V in 1914 at the beginning of the First World War, in recognition of his great work and as an encouragement to do even more in the future.

I was wearing a light-coloured evening dress with a long embroidered train, my most beautiful jewels, three ostrich feathers pinned to my hair behind my tiara and the decoration of a Dame of Honour and Devotion of the Sovereign Military Order of Malta.

In the Great Hall, Their Majesties King George V and Queen Mary were sitting on their thrones surrounded by other members of the Royal Family, including HRH the Prince of Wales (the future King Edward VIII), HRH the Duke of York (the future King George VI) with his wife Elizabeth, Duchess of York, and Their Highnesses the Duke of Gloucester and the Duke of Kent. This royal circle possessed the imperial splendour of a bygone era.

The ladies were presented individually and came forward one by one, curtseying three times to Their Majesties. All around the guests made a splendid tableau. On one side stood the ladies in their light-coloured dresses, glittering with jewels; on the other were the gentlemen in their colourful evening dress, all covered in decorations: the dignitaries in court dress, the Ambassadors with their felucas, the Officers in their dazzling dress uniforms with gold braid, the Scottish Lords very elegant in their kilts, worn with black velvet jackets, lace jabots and diamond buttons and last but not least, a figure straight out of “A Thousand and

One Nights", the Maharaja of Patiala. Tall and very good looking, he was dressed in light blue embroidered Indian silk and covered in fabulous jewels; on his sky-blue turban was a plume of feathers held by an enormous, famous diamond.

The impression of that evening was one of dazzling luxury. Looking back, it feels like a dream.

## THE WEDDING OF THE DUKE AND DUCHESS OF KENT

Guglielmo and I had the great honour of attending the marriage of His Royal Highness the Duke of Kent and Princess Marina, the daughter of Prince Nicholas of Greece and Princess Olga of Russia in the Gothic church of St. Margaret's, Westminster.

It was a memorable wedding which aroused interest all over Europe. The Duke of Kent, the youngest son of King George V and Queen Mary, was the darling of the Court of St. James. Princess Marina was a dream of beauty and charm.

Almost all the crowned heads of Europe and all the British Royal Family were present at the ceremony. The Peers sat in a semi-circle in their crimson robes edged with ermine with the Ladies beside them sparkling in their family jewels. Many members of the British Government and the aristocracy were also present. At the end of the ceremony, the splendid royal cortege, led by the handsome Prince and his lovely bride--so good-looking and so much in love--proceeded down the central nave of this historic church.

It was 1933; Britain was still at the height of its imperial power. At that moment nobody would ever have thought that Europe would be engulfed by a Second World War or that the happy bridegroom, the Duke of Kent, would be killed in an aeroplane accident.

# THE LONDON SEASON
# COUNTRY HOUSES

High society in Britain was very exclusive and not too welcoming to foreigners. They only accepted a select number of Ambassadors to the Court of St. James plus very few others. Guglielmo and I had the honour to be part of this exclusive circle.

The British thought very highly of my husband and felt great admiration for him. They bestowed honours worthy of a king on him. More than once I heard someone say, talking about Guglielmo: "He is one of us", and it was certainly not just because his mother was British. They recognised in him the personal qualities and life-style which are the mark of a true gentleman. As for myself, I was treated with friendliness and sincere affection. I appreciated these feelings very much and have always reciprocated them. The British when they give their friendship are very loyal friends. I shall never forget the warm welcomes and the kindness of my dear British friends.

In the years before the Second World War the English thought of themselves as the rulers of the world. The aristocracy still had great wealth and led a very enjoyable life. The London Season was a succession of splendid balls and receptions which took place in magnificent great houses. We received many invitations and accepted them with pleasure. Guglielmo recognized the great importance of tradition. He wore his numerous decorations on his morning coat and was proud to see me dressed in a beautiful décolleté evening gown with a tiara, jewels and the decoration of the Dame of Honour and Devotion of the Sovereign Military Order of Malta.

There were some famous beauties who drew all eyes at those elegant parties: Lady Diana Manners, the Duchess of Rutland, the Duchess of Sutherland and Lady Londonderry with her fabulous rubies. One evening Lord and Lady Londonderry gave a great ball. I remember them standing together at the top of the great staircase at Londonderry house in Mayfair ceremoniously receiving their guests: the Prince of Wales, the Duke and Duchess of York, the Duke and Duchess of Gloucester, the Duke of Kent and the beautiful Duchess of Kent, the Duke and Duchess of Athlone, the Duke of Norfolk, the Duke of Westminster, the Duke and Duchess of Portland, Lord Cecil, Lord and Lady Airlie, Lord Howard, the Lord High Chamberlain, Lord Cromer, Lord and Lady Plunket, as well as many other members of high society, the famous "London Society".

Another unforgettable hostess was Lady Cunard whose house was in Grosvenor Square. No longer young, but still very witty and amusing, she gave lunch and dinner parties in her beautiful house where one would often find: Government Ministers; diplomats, like the Italian Ambassador, Dino Grandi; artists, like Sir Thomas Beecham; and the élite of London Society. Seated at the head of the table, she was the center of attention; with her intelligence, loquacity and subtle wit, she always knew how to make the conversation interesting, amusing, and vivacious.

On the weekends, we were often invited to the great houses in the country belonging to members of the aristocracy and the descendants of the grand old families who enjoyed entertaining their friends in their country houses surrounded by parkland and gardens. Their lifestyle there was extraordinary, with dozens of servants. At the head of the servants there was always a distinguished-looking butler who wore a morning coat all day until the evening when he appeared in white tie and tails. He was the pivot around which the entire complex organization of the household revolved.

The houseparties regularly included twenty or thirty guests who drove down from London on Saturday afternoon by car with a chauffeur and a lady's maid. Sunday would pass very quickly. The British hosts and hostesses were famous for their wonderful hospitality. They entertained their guests kindly and discreetly, leaving them free to devote themselves to their favourite pastimes or sport. Those who wanted to ride a horse could go for long rides in the lovely countryside, while those who preferred golf or tennis could play on beautiful golf courses and tennis courts. Guglielmo and I loved to go for walks with our friends in the enchantingly beautiful woods and parkland or we would wander in the gardens admiring the different flowers grown with care by dedicated gardeners.

At five o'clock in the afternoon we would return to the house to meet our friends around the tea-table; if the weather was cold we would chat happily in front of a roaring fire. During the long rainy days those who enjoyed playing bridge could sit down around the card table while others were attracted by the library which was full of valuable books bound in marocco leather to please every interest and taste in literature.

In the evening, at dinner, the ladies would wear long décolleté dresses and splendid jewels while the men were in white tie and tails or dinner jackets. We would sit around a long table decorated with fresh flowers, laid with priceless porcelain and that beautiful antique silver which is the boast of England. The conversation was interesting and amusing, enlivened by the witticisms of the typical British sense of humour.

We spent many happy weekends in those great country houses together with famous people who are now a part of history: Sir Winston Churchill, Sir Austin Chamberlain, Ramsay MacDonald, Lord Reading, Lady Astor, George Bernard Shaw, the Aga Khan and many members of the British Royal Family.

The Duke and Duchess of York, the future King George VI and Queen Elizabeth, often spent the weekend at Polsden Lacey in Surrey not very far from London, as the guests of Mr. Ronald Greville.

Margaret Greville was a vivacious and witty lady. She enjoyed inviting the most important personages of politics, art and culture as well as members of the British aristocracy to her beautiful country house; she always entertained her guests with charm and intelligence; she had a circle of close friends and although it was not easy to win her friendship she was very fond of Guglielmo and myself and treated us with great kindness. She was anxious for us to meet the most interesting people who were in London at the time.

For my husband these receptions and house parties were a pleasant distraction from his work which always kept him very busy. I have vivid and happy memories of Polsden Lacey.

We were great friends of the Duke and Duchess of Marlborough who often invited us to spend the weekend at Blenheim Palace, one of the most famous country houses in England. Built at the beginning of the 18th Century by the architect Vanbrugh, it is an impressive building, surrounded by a huge park and containing an extraordinary art collection. Queen Anne ordered it to be built as a gift to Lord John Spencer Churchill on whom she also conferred the title of Duke of Marlborough in gratitude for the many victories he had won in battle, including that of Blenheim.

Guglielmo admired Blenheim Palace with its baroque style of architecture very much. Standing in front of the house he said to me: "It reminds me of the beauty and magnificence of Palazzo Barberini in Rome".

The Duke and Duchess of Marlborough used to entertain their guests very ceremoniously. At dinner time the guests, in evening dress, would walk in a procession through the reception rooms, the most famous of which was the one containing the collection of relics and flags that the great Duke of Marlborough had brought back from the battle of Blenheim.

At the end of the evening we went up the great marble staircase that led to the first floor where everyone would retire to their respective suites. The palace was very cold in winter and we were really glad to get into our enormous room which was well heated by the fire burning in the huge fireplace.

Among the usual guests were Lord and Lady Londonderry, the Duke and Duchess of Sutherland, the Duke of Alba, the Duke and Duchess of Portland, the beautiful Lady Diana Manners with her husband Duff Cooper, the future Minister, Lord and Lady Wimborne, Lady Curzon, Lady Carlisle, Lord Reading and many others.

Guglielmo and I had the pleasure of dining with Sir Winston Churchill and his wife Lady Clementine at Blenheim in 1931. Churchill felt at home there and you could tell this from his good mood. In fact, it was the house where his ancestors had lived and where he himself was born. He and the Duke of Marlborough, being cousins, were on very familiar terms, joking with one other and making witty remarks.

We met Churchill on several other occasions. We had the opportunity to get to know him well in June, 1934 when Guglielmo and I were invited for the weekend by Lord Fitzalan Howard to his house, White Lodge, in Windsor Park. In fact, Winston Churchill and his wife were also among the guests.

He was a man with a great personality; he could dominate everyone who came near him or talked to him. Guglielmo had this same gift; maybe this was the reason that when they met they understood and respected each other.

During that weekend at White Lodge, Churchill and Guglielmo talked for a long time together. To make it easier for them to talk, Lady Fitzalan Howard organised things so that at lunch on Sunday they were sitting at the corner of the table, next to each other. This was very rare; an exception to the strict British protocol.

At that time Winston Churchill still looked youthful and not too stout. He had a pink complexion, bright eyes and very small hands and feet. He had the manners of a true descendant of the Duke of Marlborough. I can still see him and hear his unmistakable, slightly nasal voice.

The Sunday of that weekend in June was very hot. It was exceptionally fine weather for England. After breakfast, at about 10, all the guests were sitting in comfortable arm-chairs placed in a circle on the grass under the ancient trees of Richmond Park. Winston Churchill was wearing soft brown leather shoes. He had his famous cigar in his mouth and a glass of whisky in his hand. It seemed as if he was holding court. Guglielmo was sitting beside him and they were talking to each other.

The bystanders never tired of listening to those two exceptional men, both so youthful in spirit, telling each other about their own unique experiences.

We were all in a good mood, happy and carefree; nobody thought that another war would break out and the subject was not even mentioned. We had put aside any problems and worries and we just wanted to relax and enjoy two days in the country with our friends Lord and Lady Fitzalan Howard.

I liked Lady Clementine Churchill very much. She was beautiful, spontaneous and showed great intelligence in her love for her husband. Few people could understand her better than I. In fact, I too had chosen an exceptional man as my husband and I knew that to be close to a person of such brilliance required great understanding and tact.

We, in our turn, were happy to return our friends' invitations and we often gave elegant dinners at the Savoy Hotel and sometimes at the Ritz in London.

# THE MARCONI COMPANY LONDON AND CHELMSFORD

I spent many years in London with Guglielmo. They were happy and busy years; we did not have a house of our own and we were always on the move. In fact, whenever there was an important assembly at the Italian Senate we had to catch the train for Rome; at other times we set sail aboard the *Elettra* towards new scientific discoveries.

We usually arrived in London by train and stayed at the Savoy hotel in the Strand, just half a mile from Marconi House, the headquarters of the Marconi Company where Guglielmo, as the Chairman of the company that he himself had founded in 1897, went twice a day. In this way he could keep his appointments with the people he wanted to meet and at the same time go on with his experiments; he had in fact located different places around London which were suitable for his research. These sites are commemorated today with little bronze plaques.

Marconi House in the Strand had been the London headquarters of the Company since May 1912. The building had previously been occupied by the New Gaiety Restaurant and luxury flats and Guglielmo told me that the work of conversion was completed in just ten weeks--an amazingly short time. I remember that the richly ornamented molded plaster ceilings and friezes were particularly beautiful. The main waiting hall on the ground floor was panelled in mahogany. The grand staircase led from the waiting hall to all the floors and there were fine stained glass windows on each landing. Under the window on the first landing were written Puck's words from *A Midsummer Night's Dream*: "I'll put a girdle round about the earth in forty minutes" to which Guglielmo had retorted: "I'll do it much quicker than that!"

My husband had a strong sense of duty. He worked very hard and for long hours because he wanted to contribute with his scientific knowledge to the success of the company which bore his name and of which he was Chairman until the end of his life. He did not think of himself just as an inventor but was proud to have been such a successful Director of the Marconi Company.

I often went with him to Marconi House. How many familiar faces come to my mind, now sadly no more! I often met Mr George Kemp. He was especially devoted to my husband, having been his most trusted assistant; he had worked for him for over thirty years. There was always affection and admiration in his eyes when he looked at Guglielmo. Mr Kemp, knowing how happy we were, said to me one day: "Marconi still has many discoveries to make; he has such a creative mind that he is certain to invent many more things for the good of humanity; I am glad he has found peace and happiness with you".

I often went with Guglielmo to Chelmsford in Essex where the wireless equipment that my husband needed for his experiments was built. Chelmsford was an important base for him. In 1898 the Company took over a building in Hall Street which was to become the first wireless factory in the world. Chelmsford was ideal as it was only about thirty five miles from London and the country around it was quite flat and therefore suitable for wireless experiments. It was also near the Port of London with its huge volume of shipping and at that time it was in ship to shore and ship to ship communication that Marconi foresaw the best chances for the future commercial success of the Company.

In 1912 the Hall Street Works were replaced by a new modern factory in New Street. Like the conversion of Marconi House, the building of the new Works was carried out in record time. The same architects, Messrs. Dunn and Watson were responsible for the designs. Building began in February and on 22 June 1912 delegates to the International Radiotelegraphic Conference in London toured the new Works which were equipped with the most up to date tools, apparatus and laboratories.

In February, 1920 the first wireless telephony news broadcast in the world was transmitted from Chelmsford. But the great event which showed the future possibilities of entertainment broadcasting took place on a summer evening in 1920. In fact, on 15th June, 1920, Dame Nellie Melba, the famous Australian prima donna gave a concert, sponsored by the *Daily Mail*, which was broadcast by wireless telephony from the Chelmsford Works. When she was shown the transmitting equipment and it was explained to her that her voice would be broadcast from the wires at the top of the towering antenna masts Dame Nellie exclaimed to the engineer who was showing her around: "Young man, if you think I am going to climb up there you are greatly mistaken"! Reassured on this point, she sang the "Addio" from *La Boheme,* "Home Sweet Home", "Nymphes et Sylvains", Chant Venitien", ending with the National Anthem. The concert was a triumph and made headline news. Letters of congratulation from listeners who had been fortunate enough to hear Melba's magnificent soprano voice through their receivers arrived from all over Europe and even from as far away as St. John's, Newfoundland.

In July 1920 a delegation to a conference of the Imperial Press Union sailed from Liverpool aboard the *Victorian* bound for Ottawa. The Press delegates received morning and evening issues of a special newspaper *The North Atlantic Times*. The Marconi station at Poldhu transmitted the telephony news service to the ship while concerts were transmitted from Chelmsford and reproduced by loud-speaking telephones. The concerts were still being received when the *Victorian* was over 2,000 miles from Chelmsford.

Demonstration news broadcasts from Chelmsford to newspaper offices in both Britain and Scandinavia continued during July and August 1920 but the British Post Office had only granted the Marconi license for experimental purposes and no normal program of entertainment could be transmitted. Suddenly, the Post Office withdrew the Marconi license and from that time no professional transmissions were permitted in Britain. It was not until January 1922 that the Postmaster General finally authorized the Marconi Company to set up a station. Transmission time could not exceed half an hour per week!

A transmitting station was set up in a wooden hut at Writtle near Chelmsford. The weekly half-hour was scheduled for Tuesday evenings between 8 p.m. and 8:30 p.m. and the first radio entertainment broadcast went out on the air on 14th February 1922. The call sign was 2-MT (Two-Emma-Toc). The broadcaster in charge of the programmes was P.P. Eckersly who delighted his listeners with his spontaneous humour and informal approach. He later became the first Chief Engineer of the BBC.

Not long after 2-MT began its weekly half-hour broadcasts the G.P.O. authorised the Marconi Company to set up an experimental station at Marconi House in London. It could transmit for a maximum of one hour per day. The transmitter and studio were in a room at the top of Marconi House and began transmission on 11th May 1922. The call sign was 2LO. The man in charge of the programmes was A.R. Burrows who was to become the famous "Uncle Arthur" of the BBC.

The inauguration of broadcasting in Britain by the Marconi Company naturally brought about a great number of applications for permits from other companies. In October, 1922 six of the most important manufacturers, Marconi's Wireless Telegraph Company, the General Electric Company, the Metropolitan Vickers Company, the Western Electric Company, the British Thomson-Houston Company and the Radio Communication Company set up the British Broadcasting Company Ltd.

On 14th November, 1922 the British Broadcasting Company took over the evening transmissions from the London station 2LO. In April 1923 the BBC acquired the building at No.2, Savoy Hill and studios were built there to take over from those at Marconi House. On 31st December, 1926, with the granting of a Royal Charter, the British Broadcasting Corporation took the place of the British Broadcasting Company with Sir John Reith as the Director General. In the four years that the "Big Six" companies had been directly responsible for broadcasting in Britain listeners had grown from a few thousand to over two millions.

When we were in London we often used to visit the BBC, of which Guglielmo was, as I have said, a founder member and which was not far from the Savoy Hotel at Savoy Hill. We listened to the broadcasts from the BBC's main radio set; then Guglielmo had discussions with the head technicians and gave them advice on how to improve the quality of the sound. I have pleasant memories of the friendly terms we were on with the personnel of the BBC.

In the spring we always went to Chelmsford for the sports day organized by the Marconi Company. It was a festive occasion when all the employees, the veterans and the directors of the company got together. There were races and games outdoors for the children and young people. I still have a photograph which shows Guglielmo sitting beside me with our daughter on his knee. He is smiling and pointing to some children who are taking part in the sack race. Little Elettra was then four years old. She had been asked to join in but while she was racing in front of us she suddenly jumped out of her sack and took refuge on her father's lap where she stayed for the whole time, to the amusement of those present. Elettra has never forgotten this episode and laughs about it now.

My daughter and I are often invited by the Marconi Company to commemorate my husband and we are present at the unveiling of memorial plaques at the celebrations in his honour. They usually end with dinners and speeches made by the Chairman and myself in which Guglielmo's great works are remembered.

# THE INAUGURATION OF VATICAN RADIO 12 February 1931

Pope Pius XII before becoming Pope in 1939 was already a close friend of my mother's family and ever since he was a young man he and my father, Count Francesco Bezzi Scali, had been good friends. In May, 1911 they had gone together as members of the Papal Delegation that went to London as guests of the Duke of Norfolk at Norfolk House for the Coronation of His Majesty King George V. It was the first time since the days of Henry VIII in the XVI century that a Papal Delegation had taken part in the Coronation of an English sovereign. His Holiness Pope Pius XII knew me well; before becoming Papal Nunzio in Bavaria he had given me catechism lessons when I was still a young girl. As I have already mentioned, it was through me that he got to know Guglielmo personally and once he was sure of the depth of our feelings he gave his blessing to our marriage.

It was also through me that Guglielmo became friendly with His Holiness, Pius XI. The Pope was also a man of science and took a great interest in Guglielmo Marconi's research and discoveries. When we were in Rome the Pope was always happy to receive us in private audience. These visits went on and on; we used to come out of His Holiness's study after two or three hours, sometimes at three in the afternoon to the dismay of the gentlemen of the court and the various prelates in waiting who were hungry for their lunch. It was wonderful to listen to the conversation of those two men as they talked quietly and intensely about the most lofty scientific questions.

I shall never forget an audience which took place in 1930, immediately after the Reconciliation, when Pius XI in a firm and decided voice asked Guglielmo to design and personally set up a powerful modern radio station in the Vatican so that the Pope's voice could be heard all over the world and make it possible for everyone to receive his apostolic blessing "Urbi et Orbi". Guglielmo accepted without a moment's hesitation. He said he was delighted and honoured to work on such an exceptional project. He set to work with the greatest enthusiasm. It was a project that was really important to him. I was happy to think that thanks to my contacts with the Vatican and the Holy See Guglielmo by marrying me had also become part of that world and had been entrusted with the task of building a superb new radio station which still gives pleasure to people all over the world. During the following six or seven months we often travelled from London to Rome so that my husband could personally oversee the building of the Vatican radio station.

The inauguration ceremony took place on 12th February 1931 in the radio station in the Vatican gardens. Guglielmo and I arrived there at around half past eleven and he immediately started to check that everything was working perfectly. In the meantime a little crowd of guests was beginning to collect, composed mainly of Cardinals, important pontifical dignitaries and some members of the Roman aristocracy, including my Uncle Prince Luigi Barberini with his sons Enrico and Urbano. His Holiness Pope Pius XI arrived at around twenty past four and was received by Guglielmo and myself with a few others. Accompanying the Pope were the Secretary of State, Cardinal Eugenio Pacelli the future Pius XII, Monsignor Alfredo Ottaviani and Father Gianfranceschi, who had been appointed Director of the Vatican radio. There were also Prince Marcantonio Colonna, my cousin Marquis Giovanni Battista Sacchetti, Prince Leone Massimo and Marquis Giacomo Serlupi Crescenzi. The commanding officer of the Papal Guard Prince Aldobrandini was represented by my father Count Francesco Bezzi Scali who was then Brigadier General of the Papal Guard.

My husband showed the Holy Father around the radio station. He took a great interest in the equipment and then both stopped in front of the microphone. Guglielmo was very pleased with the outcome of his work which he now gave to the Pope and to the Church. I can still hear his words: "The use of microwaves in permanent operation is being demonstrated here today for the first time in the world. I wish to express my deep gratitude to your Holiness for this and also for giving me permission to carry out a series of experiments here. My happiness as an Italian and the honour that is mine today as a researcher are crowned by the thought that this first application of an important scientific discovery is being carried out not only by the will and in the presence of a great Pope but through the skies and land of Italy on the anniversary of a happy day for our country and for all humanity, with the aim of making communications easier between the nations and contributing to bring true Christian peace to all the peoples of the world. This work I have accomplished fills my heart with pride and hope for the future as an Italian and as a scientist. May my work help to bring peace to the world."

After a brief pause, he added with feeling: "I have the great honour to announce that in a few moments his Holiness the Pope, Pius XI will inaugurate the Vatican City radio station. The electric waves will carry his words of peace and blessing through space to the whole world. For almost twenty centuries the teachings of the Pope have been carried all over the world but this is the first time that the Pope's own voice can be heard simultaneously in the four corners of the Earth. With the help of God who has put so many mysterious forces of nature at the disposal of humanity I have been able to prepare this instrument which will give the faithful all over the world the consolation of hearing the voice of the Holy Father. Blessed Father, the work that your Holiness entrusted me with, I now consign to you: its accomplishment is consecrated today by your presence; Holy Father, let the world hear your noble words." Then my husband was silent.

After a few moments of silence the voice of Pius XI was heard delivering his first radio-message to the world in Latin. He began solemnly, in biblical tones, with these words: "Hear, Oh Skies, that which I have to say; let the Earth hear the words of my mouth... Hear and listen, Oh far away peoples". Then, turning to Guglielmo, the Pope said: "Well, Marquis, we asked you to give us some idea, some demonstration of how and by what scientific means you find the path of these waves that nobody sees and nobody hears. All the same, we are still curious, we still want to know how the human mind can see as it were a "vision" that is so distinct, how it can measure so exactly what the eye cannot see and the hand cannot touch. We can do no more than repeat our congratulations for that which the divine goodness, the divine power has allowed you to achieve in order that the secrets of the divine omnipotence, the divine wisdom that so admirably rules over all things should be of real benefit for humanity. Benefits from on high in the widest applications that your heart and your mind can devise".

Pope Pius XI thus expressed his satisfaction and his gratitude to the creator of the new broadcasting station, my husband Guglielmo Marconi. For the first time in the history of the Church the voice of a Pope had been transmitted and recorded. At the end of his speech His Holiness gave his blessing "Urbi et Orbi".

# THE RITZ HOTEL
# London

On 30th September, 1967, I read in the Roman newspaper "Il Tempo" about the death, the day before in Paris, of Prince Felix Jusupov, who was a cousin by marriage of the last Tzar of Russia, having married the Grand Duchess Kira.

When we were in London Guglielmo and I often went to lunch in the restaurant of the Ritz Hotel in Piccadilly. It was a smart meeting place. During Ascot Week, in the full swing of the London Season one could meet the cream of British high society and many celebrities from all over the world: Royal Princes, Maharajas, Lords, Ministers of the British Government--Winston Churchill was often to be seen--and a great many beautiful ladies.

In the spring of 1928, we went, like every other year, to the races at Ascot where we had been invited to the Royal Enclosure. As we entered the lobby of the Ritz, we met two very smart young gentlemen: the Grand Duke Dimitri of Russia and Prince Felix Jusupov himself. Prince Felix, who usually lived in Paris, knew Guglielmo well. He came up to us to greet him and compliment him on how well he was looking. "You look so happy," he said. He was very charming to me. We decided to have lunch together at the same table.

The two Russian Princes were both very charming men. While the Grand Duke Dimitri was calm and rather silent, Prince Felix was lively, exuberant and spoke with great verve. Guglielmo asked him to tell me exactly what had happened at the assassination of Rasputin, the diabolical monk who had had a fatal influence on the beautiful and unhappy Alexandra, the last Tzarina of Russia, convincing her that he was able to heal her son, the little Tzarevic Alexander, who suffered from the terrible disease haemophilia, inherited from his mother.

Prince Jusopov consented; he spoke in English, moving his hands with graceful gestures as he spoke. His lively black eyes shone in his handsome oval face. He told us that in spite of the strong poison Rasputin had drunk and the revolver shots in his chest, he did not die but on the contrary continued to struggle.

"Then Grand Duke Dimitri and I dragged him to the River Neva. We stabbed him again and again with our daggers. All around him the water had become completely red. Nevertheless, the tortured flesh of that devil continued to writhe. Finally, he died."

Remembering the tragedy, Prince Felix still showed by the look in his eyes the horror he had felt on seeing all that blood gushing from Rasputin's body and making the waters of the Neva run red. I found his description of that frightful episode quite horrifying; Guglielmo instead looked with admiration at this young Russian Prince who had shown such boldness and courage, risking his own life to destroy that evil man who had had such a negative influence on Russian history.

# RUSSIAN ARTISTS
## Covent Garden

During the London Season of 1928, on a wonderful evening at the opera at Covent Garden, Guglielmo and I had the pleasure of listening to the great Russian singer, Ilodor Saliapin.

He was a handsome man: tall, charming and youthful looking. Guglielmo and I knew him well because, like us, he was staying at the Savoy Hotel and we sometimes dined together. He spoke Italian and we had very entertaining conversations with him. He was at the height of his career and was much in demand by the most prestigious theatres throughout the world, including the Metropolitan Opera House in New York and, of course, Covent Garden in London. He loved Italy where he sang and met with great success at the Theatre of the Opera in Rome and La Scala in Milan.

A few evenings later, again at Covent Garden, Guglielmo and I watched a wonderful performance of Russian ballet. After the performance, we were invited to a comfortable little sitting room behind the scenes to meet the impresario who had created the ballet company, the famous Sergei Pavlovic Djagilev--a most talented and original man. He had a strong personality, great artistic talent and exceptional intuition. He was very famous in the world of dance because of the new interpretation he managed to give to his art; effectively re-inventing modern ballet. Furthermore, he himself created the magnificent stage set designs for his ballets and also occasionally for Saliapin's performances.

I was wearing a beautiful rose-pink embroidered evening dress by Ventura. Diagilev, a connoisseur of life and beauty, looked at me admiringly and said to my husband: “I have never seen such a beautiful woman.” Guglielmo smiled at him with pleasure while I was left feeling a little self-conscious by the compliment.

The atmosphere that evening was light-hearted and we were all in high spirits. It was a really unforgettable meeting with that Russian genius.

# ELETTRA AS A LITTLE GIRL

One day I was at Forte dei Marmi, waiting for the return of my daughter Elettra who was out sailing on the yacht with her great friends, Prince Corsini, his sister Donna Anna and his son and daughters. The sea was calm, although a strong westerly wind was blowing but Elettra is certainly not afraid of the wind because she really loves the sea and sailing which she has inherited from her father.

In that moment I thought of Guglielmo, who told me about the time when he was fourteen and was given a little sailing boat as a present by his father Giuseppe, while they were on holiday one summer in Livorno. Encouraged by his love for the sea he had no fear of sailing out in the open sea, defying the strong gusts of wind. Sometimes his parents were worried because he was late in returning; then his father commented: "One of these days all we are going to find of Guglielmo will be his cap on the beach."

Our daughter Elettra went on board the yacht bearing the same name when she was just a few months old, accompanied by her nurse, Vincenza Pitocco, a beautiful young woman from the Roman countryside. She was nineteen years old, intelligent and affectionate, with big shining dark eyes. My husband did not want to leave the baby in her cot, fearing that the rolling of the yacht could make her fall out. So he prepared a little bed for her in the half-open bottom drawer of a wardrobe in her cabin.

The baby was our pride and joy and was great company for us. When Guglielmo took a break from his experiments he came up onto the deck of the yacht and sat down in an armchair in front of the playpen where Elettra was playing in the shade of a long awning. He smiled at his little daughter who greeted him with screams of happiness. How they enjoyed being together! They got on so well because they were so much alike.

Elettra took her first steps on board the yacht. It was very difficult for her; she lost her balance because of the movement of the yacht. How lovingly her father helped her, holding her steady! It was delightful to watch them.

As our daughter grew up Guglielmo took an affectionate interest in her physical and moral development. He had strong principles and he wanted the little girl to be well brought up and keep her character unchanged. Elettra has always had an enthusiastic and cheerful nature. Even when she was a little girl she always wanted to take an active part in what was going on around her.

One day at the beginning of June, sailing towards Liguria, we anchored in front of Porto Santo Stefano. In those days the Argentario with its characteristic watch-towers along the coast had not yet been discovered by tourism but was a lonely place. Taking advantage of the sun and the calm sea we went away from the yacht in a small rowing boat. Guglielmo was at the oars and I was holding Elettra, dressed in a light sleeveless sundress. She was ten months old and it was the first time we had taken her with us in a small boat. When we got to the open sea away from the yacht I decided to dive into the water so I entrusted the baby to Guglielmo who left the oars.

While I was swimming near the rowing boat my husband had the idea of undressing Elettra and lowering her, holding her tightly under her little arms, into the sea towards me. We were both curious to see how the little girl would react: she began to kick her legs, screaming with joy, and she cried only when her father took her back into the boat, holding her tightly in his arms. We were very happy to see that Elettra had our same love for the sea. Very soon she showed us that it was more natural for her to live on the sea than on the land. In fact, she was happy to sail on board the yacht and considered the Elettra her real home. I remember that one afternoon, going back by motorcar to Porto Santo Stefano after having spent a few days in Rome, we saw the yacht anchored off the port waiting for us to embark. Guglielmo asked his daughter: "Whose is the *Elettra*?" "The *Elettra* is Elettra's!" she answered.

All the yacht's crew came from the province of Liguria, with one exception; a sailor from Tuscany. He was called Agostino and he was the son of Vincenzo Polacci, the lifeguard who taught my brother Antonio and myself to swim in Forte dei Marmi when we were children. Soon after my marriage with Guglielmo I remembered him and I arranged for him to come on board as the sailor at our personal service. Agostino was very fond of our little daughter Elettra and he showed her how to wash the deck and how to polish the brass fittings of the yacht. She enjoyed helping him.

When Guglielmo died so suddenly, life on board the *Elettra* ended too. Agostino went back to Forte dei Marmi where he and his wife Marianna opened a beach Lido which he called "Bagno Marconi"-"Marconi Beach". He called his son Guglielmo and his grand-daughter Maria Cristina. All his life he remained most devoted and attached to us--feelings which we sincerely reciprocated.

When they learned of the death of the Marchesa Marconi all the people on holiday in Forte dei Marmi saw with emotion the flag of "Bagno Marconi" at half-mast in sign of mourning.

One summer day when Elettra was already four years old we were sailing in the Mediterranean while a strong south wind was blowing. I was sitting in a wicker chair on the deck of the yacht, holding the little girl tightly in my arms. As a precaution, we were tied down with ropes; an original idea of Guglielmo's so that we could enjoy the storm without the danger of falling into the water. The yacht was rising and falling; when it rose the waves fell on us with a blow, wetting us completely. Elettra was happy, laughing and shouting, "How many splashes! How many splashes!" We did not stay there long because Guglielmo and the Captain were rather worried and made us take shelter inside the yacht.

I should like to record in full an amusing article written at the time by the journalist, Ettore Fabietti:

"In the year 1927 Marconi married in Rome and as soon as possible he took his young bride on board the *Elettra* where the couple made their home. The Marchesa Maria Cristina is the hostess of this splendid floating home. With her radiant smile and all the charm and grace of a great Roman lady she welcomes the famous guests who come on board. She loves music, literature, history of art; she can carry on a conversation that is both witty and practical with the learned foreigners who climb on to the deck of the Elettra to pay homage to her husband and are so fascinated by them both that they can hardly bring themselves to say goodbye.

Since 1930 a very small person has dominated everyone on board, beginning from Marconi and his wife, who satisfy her every whim and run to do her bidding. The little girl is called Elettra, like the yacht, and is the most recent flower to blossom from the father's vigorous stock. Queen Elena is her godmother. Elettra seems to be the visible spirit of the beautiful yacht.

She gets up early every morning and rushes up on deck, just at the time that the sailors are washing it. She wants to play, jumping in the rivulets of water trhat run on the smooth foredeck. She only obeys the Captain; and do you know why? Because, ever since they gave her a smart yachtsman's uniform, she considers herself a real sailor and the Captain has explained to her that being a sailor she owes him absolute obedience like the officers and the crew. When he whistles she stops playing and runs up to him. Standing to attention, she raises her little hand in a smart salute: "Aye Aye Sir!" But she says that one day she will command the *Elettra*.

She adores her father and his face lights up with joy whenever he sees his charming little girl. She treats him very directly without any shyness and in her blue eyes he sees those of his beloved wife.

During the holidays, when the white yacht lies at anchor in the bay of Santa Margherita Ligure for a few months, the little girl's joy is at its height. Daddy teaches her geography which is her favourite subject although she can't pronounce the names correctly. Mummy and a sailor teach her to swim, her grandparents to imitate Shirley Temple and her governess to read and write.

Marconi usually lunches with his family at a small table on deck.

It was in this atmosphere that he carried out his new experiments on very short and micro-waves."

When her father died, Elettra was still very small but she always had vivid memories of him. Guglielmo for his part loved his daughter and was always very patient with her. He was never bored in her company. "Elettra is electric." he often said.

In fact, she was a very lively little girl and we were worried that she might slip and fall into the sea when she was running on the deck. So my husband, with his usual foresight, had a net made of rope put up along the yacht's railing as a safety precaution.

When Elettra was older and could understand, Guglielmo started to explain to her the manoeuvres to anchor the yacht: the arrival of the pilot at sea off the port and the procedures for berthing the yacht alongside the quay. He also explained to her the various ways to fish: the use of lights, nets, harpoons and hooks. Elettra listened very carefully to the explanations of her great father; she took it all in and was full of admiration for him. It was touching to see them together; Guglielmo, wearing his yachtsman's clothes and his cap with the badge of the Royal Yacht Squadron of Cowes, spoke to the little girl as if she was a grown-up. "The child is clever." he said. "One must not explain things to her in a babyish way."

Elettra still remembers that her feelings for her father were of loving friendship and respect. As I have already said, they had many characteristics in common, in particular a sweet nature together with a strong character. They understood each other with just a look. "Elettra and I are linked by radio!" he told me more than once. It is true. When they were together they were happy, nothing troubled them, they understood each other perfectly, laughing and joking and enjoying themselves very much.

Sometimes Guglielmo took his daughter by the hand and led her to his radio- station on board. I followed them. While he was showing his kingdom to his little girl, he pointed out the great valves behind their protective grille and said to her, "Do you see this? It's very dangerous. Don't touch." Elettra knew that when her father stayed for a long time in the radio-station it was because he was inventing great things. Talking about this, I remember an amusing episode:

One day we were in Rome, staying with my parents in Via Condotti. It was early in the morning and we were having our breakfast in our bedroom which was already lit up by the sun. From the windows one could enjoy the view of the courtyard with the trees in bloom. Suddenly Elettra came in to say good-morning to us. She must have been about four years old. She sat down on a stool beside us, holding a little dress for her doll in her hands, needlework prepared especially for her by her English nanny to teach her to sew. Evidently the little girl found it very difficult to use needle and thread because all of a sudden after a short silence she lifted her fair curly head and looking up with a questioning look in her big blue eyes she asked Guglielmo, "But Daddy, when are you going to teach me wireless sewing?" (In Italian "thread" and "wire" are the same word -"filo"). How we laughed and how Guglielmo enjoyed his little daughter's witty remark. From then on he never tired of telling the story.

Even though she did not have her father near her as she grew up, Elettra continued to resemble him more and more. Together with the lively Italian character she has the same English sense of humour. Like Guglielmo she is gifted with self-control and she is able to overcome difficult moments with calm and intelligence. At the same time she is emotional; even little things can make her happy, but she can suffer deeply, shutting herself up within herself if she is attacked unjustly.

# BLIND NAVIGATION

I should like to record the words spoken by my husband during the conference held in New York on 20 June,1922, to the members of the Association of Radio-electrical Engineers.

"In some of my tests I have noticed the effects of reflection and deflection of these waves by metallic objects miles away. It seems to me that it should be possible to design apparatus by means of which a ship could radiate or project a divergent beam of these rays in any desired direction, which rays, if coming across a metallic object, such as another steamer or ship, would be reflected back to a receiver screened from the local transmitter on the sending ship and thereby immediately reveal the presence and bearing of the other ship in fog or thick weather. One further advantage of such an arrangement would be that it would be able to give warning of the presence and bearing of ships even should these ships be unprovided with any kind of radio".

As a consequence of this research, another of the important inventions of my husband, Guglielmo Marconi, was what he called "blind navigation" and which after his death was called "radar" by the British who used it during the Second World War.

I have already said in these memoirs that for some time Guglielmo had been carrying out experiments on short and very short waves, concentrating especially on the so-called "beam system", on board the yacht, Elettra.

Once through the Bay of Biscay, we sailed along the French coast near the famous rocks called "Toussaint": This stretch was well-known to Guglielmo and to every sailor because of the danger of the "Needles" (high and jagged rocks like needles). Furthermore, we constantly met areas of very dense dark fog which was a great danger for all the ships in this and every sea.

It was there, in 1931, that Guglielmo decided to put his ideas on blind navigation into practice. On these occasions he was filled with an incredible enthusiasm. Being so fond of sailing, his greatest worry was fog; he was always trying to think of an as yet undiscovered way to avoid collisions between ships and to locate obstacles in the darkness. His really wonderful aim was always to work to save human lives. Therefore, Guglielmo decided to study and develop a system for sailing even in conditions of scarce visibility. In fact, he carefully continued his studies and experiments and once again reached his goal; I, as an eye-witness, can thus bear witness to Guglielmo's incredible invention of radar, which he called blind navigation. I was present once when he said in one of his speeches: "I hope that I shall soon succeed in showing how these waves will be able to help navigation, especially in fog".

All his studies on very short waves, between 1932 and 1934 were crowned with success towards the end of his life with the accomplishment of blind navigation in the Gulf of the Tigullio. He had transferred his experimental base to the Hotel Miramare in Santa Margherita Ligure in July 1932 and this was where he developed the apparatus for this incredible invention.

One summer morning Guglielmo asked me to bring him some thick white cotton sheets which, helped by Captain Stagnaro, I attached to the glass of the rectangular windows around the bridge in front of the helmsman, so that the yacht would be sailing blind. Guglielmo stood there, giving the orders simultaneously to the captain and the helmsman. I never left his side. In these moments of great emotion, my heart beat fast, while I hoped that the experiment would be successful. That was the reason Guglielmo and I were so close! We were never apart. He always told me that my presence was a great help to him and that I brought him luck so that his research was crowned with success. This too is a proof of our deep, mutual love.

We sailed from Santa Margherita Ligure. Along the coast, at the entrance to the little port of Sestri Levante, two buoys were placed the exact distance from one another, so that the bow of the yacht Elettra could pass between them. On the tower on top of the promontory overlooking the picturesque bay of Sestri Levante my husband had installed the apparatus that he was going to use for these experiments. From then on, the tower was called "Torre Marconi" and a memorial plaque was placed upon it in memory of Guglielmo and this event. I shall always remember the intense emotion that I felt when, for the first time, sailing blind, with the windows covered as I have described above, the bow of the Elettra slipped precisely between the two buoys. Guglielmo's expression was one of satisfaction and happiness; once again he had achieved his purpose. In fact, to the admiration and astonishment of all those present, both on board and on shore, the yacht Elettra sailed safely between the two obstacles.

My husband immediately understood the great importance of his achievement. He did not want to lose time so we went to London for him to report the result to the Marconi Company. Guglielmo said he would like to give a demonstration of his latest invention to an expert. So a British admiral, Admiral Sir Henry Jackson, who was the harbour-master of the port of London and an expert on fog came on board; fog was a phenomenon which was especially frequent in the area around the mouth of the Thames. The experiment was a complete success. Admiral Jackson stayed on board the yacht as our guest for three days. I still have some photographs which show him standing beside Guglielmo, myself, Captain Stagnaro and our officers just as the bow of the yacht passed between the two buoys. Admiral Jackson, impressed by what he had seen, immediately approved Guglielmo Marconi's new invention, which was officially recognized, patented and produced by the Marconi Company. It thus remained in England, just as the War was getting close.

Meanwhile, our dear little daughter Elettra was growing up, carefree, on board the yacht. She too was present at this event and remembers it still; she was growing all the time and she realized that she had a very special father who gave wonderful inventions to the whole world.

# THE *ELETTRA* AND SAILING IN THE ATLANTIC

In 1920 Gabriele D'Annunzio went on board the *Elettra* which was anchored in the port of Fiume and presented my husband with a photograph of himself dressed as a legionary with the following dedication:

To Guglielmo Marconi's
White yacht
that sails through miracles
and animates the silences of the air.
Gabriele D'Annunzio

Guglielmo treasured this photograph and kept it in his studio on board the yacht.

The *Elettra* was built in the Ramage and Ferguson shipyard and launched on the River Clyde in Scotland in 1911. It had been ordered by an Archduke of Hapsburg, who named it Rowenska. At the outbreak of the First World War the yacht was requisitioned by the British Admiralty and used as the flagship of the great sailing-ships called the "Drifters".

Guglielmo was proud of the origins and the history of his yacht, the *Elettra.* The British admiral who commanded her told him that he had often seen the Drifters return to port flying the flag with a white skull on a black field, the sign of victory over the enemy, from the top of the main mast.

Guglielmo loved the *Elettra* which he bought outright at the end of 1919. In fact, the yacht meant so much to him that when our daughter was born he decided, unhesitatingly and with great enthusiasm, to call her Elettra. When he looked up the "Gotha of Yachts" in the library on board he found to his satisfaction that never before had the name of a boat been given to a human being but only the other way round.

The *Elettra* was about eighty metres long, with a tonnage of eight hundred tons and could reach an average speed of twelve knots. As one of the great yachts it was insured by Lloyds at the maximum classification, "a hundred to one".

The crew was made up of twenty-five men, a captain, four officers and a telegraphist. At first the crew-members were all from Naples, commanded by a very competent naval officer, Captain Raffaele Lauro. Later, since the yacht was laid up in Genova from October to March, an excellent crew from the province of Liguria was taken on, with a first-rate captain, Captain Girolamo Stagnaro, from Camogli, who was the holder of a gold medal for gallantry.

There were two motor-boats on board for going ashore and on excursions. One was a closed saloon with window-panes like a coupé. It was very elegant with a pointed bow in polished wood like a violin. It had an Isotta Fraschini engine and was my husband's favourite. We used it mostly in the evenings when I wore a long dress and Guglielmo a dinner jacket or tails.

The second motor-boat was an open Thornicroft which was more suitable for trips when the sea was rough. We enjoyed using it for excursions along the Cornish coast, as I have already described, and also to go up the River Fowey. Near Southampton we used to sail up the River Humble; we were charmed by the silence and solitude we found there.

When we went on these excursions our helmsman was a very competent and quick-witted young fisherman from the island of Ischia, called Tommaso, who did not speak a word of English but made himself understood perfectly by the British sailors using gestures; he was enthusiastic about the great quantity of fish he managed to catch in England, especially in Cornwall.

The *Elettra,* apart from the crew, could carry nine people comfortably. The dining room, in the centre of the boat, was roomy and full of light with big windows on both sides which gave one an excellent view of the coast and the sea; an awe-inspiring sight when there was a storm out at sea. It was a cheerful, comfortable room with white lacquered walls and blue fitted carpets on the floor. There was a sofa and armchairs covered in a pale blue and white chintz which I had bought in

London at Liberty's in Regent Street; the same material was used, in different colours, in the various cabins. A big rectangular Chippendale table and a fireplace with a mirror above it completed the furnishings. We used to spend a lot of time in this room.

The yacht had a excellent heating system which was used mainly when we sailed in the North Sea and the Atlantic. In warm climates, instead, even when we were inside we could breathe the sea air that came from the bridge through a large wind scoop.

In a square cabin next-door to the dining room was the wireless station: a perfectly equipped laboratory for radio-telegraphic and radio-telephonic transmission and reception. I can still see it, with the big, heavy machinery for the experiments on short waves installed along two walls at right angles. Every year the equipment increased in number and size, taking up part of a third wall. The big, powerful valves were protected by a strong metal grille because you could be electrocuted instantly just by brushing against them. They were really impressive.

How many hours I spent with Guglielmo--who always wanted me near him--in the wireless station of the *Elettra*, that fascinating corner which was his kingdom! He moved around in there calmly and happily, looking over his instruments with a bright, affectionate eye; and he smiled at me, letting me know how happy he was to have me with him. The particular way in which Guglielmo looked at things and scrutinized them--which I often see with feelings of emotion in the gaze of my beloved grandson Guglielmo, Elettra's son--almost always made it seem as if he was intent on discovering some mystery that was hidden from other people.

Whenever everything went just as he wanted, his face would light up and he would give me his sweet smile. He put his ideas into practise at once, certain of the result. His self-confidence and determination were what impressed me most. I saw the spark of genius in Guglielmo.

He often said to me: "Without the *Elettra* I would not have been able to do the necessary experiments in the Mediterranean and the Atlantic and I could not have developed and gone on with my research on very short waves. Instead, in this unique, floating laboratory, I could carry out my dreams: inventing "the Marconi beam system" and blind navigation and even discovering the system for extracting gold from sea water."

In fact, he continued his wonderful inventions on board the *Elettra* and I was an eye-witness to them: he made enormous progress in the field of radio. He considered that the yacht was indispensable for his life as a scientist. He always sailed far from land in order to carry out his experiments. With the passing of the years I had proof of all the countless inventions for the good of humanity that Guglielmo created, with me at his side, on board the *Elettra*.

He had already started his studies on the beam system in 1916 when he succeeded in communicating with a villa on a cliff overlooking the sea at Castiglioncello in Tuscany from a balcony of the Miramare Hotel in Genova. Guglielmo often spoke to me about these preliminary experiments of his on short waves which took place during the First World War. A few years later, having made sure that the beam system worked, Guglielmo, in agreement with the Marconi Company in London, presented the project of a new long-distance short wave transmitting station, called a "beam system" to the British government which immediately gave its approval to this new invention.

In 1928 we lived in Rome for a short time. One day Guglielmo held a conference, at which I was present, in the ancient Roman Augusteum, where he said the following words: "I have been mistaken. Up till now I have been working on long waves, but I must work on short waves instead".

For the short wave transmissions he used special radio antennas which he called "the towers". They were high steel towers with the top shaped like a huge hammer. He himself had designed them and supervised their construction; they had to be in an open space, far from buildings and near a high power radio station.

We often used to visit these installations in Britain and the United States; we also had the opportunity of seeing them in the great radio centre of Osaka in Japan in 1933, the year in which my husband and I went round the world together.

When Guglielmo stood at the bottom of a group of those imposing towers which seemed to challenge the skies, he looked at them with great satisfaction. I remember that he often said to me as he looked at them: "Do you see? The long distance communications leave from the top of these towers by means of electromagnetic waves and reach Australia, South Africa and the rest of the world".

Today all this seems quite natural because the "beam system" has been installed in every continent but at that time it was an astonishing thing.

Guglielmo told me that the British government was the first to understand the great importance of having the "Marconi beam system" radio stations so as to be able to communicate reliably and secretly with Great Britain's various dominions in every corner of the world.

In his floating radio station on board the *Elettra,* Guglielmo seemed to be rejuvenated. He was like a wizard among his magic spells. He never showed any sign of tiredness and worked uninterruptedly, even during the terrible storms that raged in the North Sea and the Atlantic when the yacht's bows seemed to be shattered by the crashing waves. During these storms I would stay in the dining room which was in the centre of the boat and the most comfortable place on board. Guglielmo showed his great love for me in his thoughtfulness for my comfort. He used to make me lie down on the sofa and wrap me up in a tartan rug. He did not want to see me wandering about the yacht as he was afraid that I might be swept overboard or fall and hurt myself; in fact it was really dangerous. Then, having made sure that I was safe, he would return to the radio station with the confident step of a seaman to go on with his experiments.

Guglielmo very often received signals through the headphone when there was a storm at sea. In fact, as we sailed further and further away from land in the Atlantic he could choose the position of the beam of radio waves he wanted to penetrate. Keeping his feet and holding on to any support within reach he would come and go between the radio station and the dining room. He said to me: "You know, Cristina, now we are inside, right in the middle of the beam of very short waves; the signals are loud and clear. Come and listen!" Then he took me with him into the radio station, put the headphone on me and smiling and happy as two children we listened to the transmission that came through clearly without interference.

Once again the miracle had happened. Once again every difficulty had been overcome. But how many days had been spent studying and experimenting in order to achieve the successful results that he always foresaw.

In order to enter the beam zone where the signals were strongest we had to sail out into the ocean far from Britain, Spain and Portugal. The *Elettra* with Guglielmo Marconi in command could face any storm. He and Captain Stagnaro who was also a first-rate seaman were always in agreement. The waves were as high as mountains; we climbed them and then seemed to fall into an abyss, among the tossing waves. We felt very small in the immensity of the Atlantic Ocean. It was really terrifying! And so with courage and strength of mind we bore the rolling, the pitching, the tail sea and the tire a bouchon. Guglielmo, being an expert sailor, could distinguish and give the correct name to the various changes that the sea underwent according to the strength and direction of the wind.

One characteristic of Guglielmo Marconi was his great physical strength, together with his exceptional tenacity in continuing his studies and research during the violent storms at sea. In those hours of creative work he was happy and full of enthusiasm. I understood how important it was for him to have no worries so that his genius could reach its maximum expression. He was grateful to me for my thoughtfulness and I could feel that he loved me deeply and I understood why he often told me that our great love and the happiness and encouragement I gave him helped him to develop his insights to the full.

When, in order to complete his experiments and discoveries, Guglielmo felt that it was indispensable to continue sailing out in the ocean, he did not allow the yacht to change course and make for a port to take shelter but expected to continue the voyage for days and nights on end. He was never sea-sick and I can say the same of myself. Sometimes the crew showed signs of tiredness because it was really very difficult to sail for a long time on board a yacht (even one of 800 tons) in a stormy sea, under torrential rain and in gale-force winds. But Guglielmo was able to transmit his incredible energy to everyone. To overcome these difficulties he always wanted a first-class captain in command of the yacht. During the very bad storms it was impossible for the ship's chef to do any cooking so one ate sandwiches and drank sherry or cognac.

Although I, like Guglielmo, was a good sailor, the crash of the waves against the hull made me feel really ill; however, I never complained so as not to take my husband away from his ceaseless work. I always lost a lot of weight during those long, tiring voyages. Guglielmo instead was the only person on board the *Elettra* who did not show any signs of being tired and whose energy never flagged. In fact he was very disapproving if the ship's cook or other crew members were sea-sick.

Unfortunately, I cannot describe everything, but there is so much I still have to tell about Guglielmo Marconi--about our long voyages in the immense Atlantic and in the Mediterranean, "mare nostrum" as it has always been known. And about his discoveries and important experiments which continued right up to the end of his life.

Quite often, around the end of March, from the port of Genova, and more precisely from the little Duca degli Abruzzi harbour in front of the Italian Yacht Club, we used to go aboard the *Elettra* and set off for a long voyage towards the Gulf of Lyon and the coasts of Spain and Andalusia.

We would sail along the coast, back and forth through the Straits of Gibraltar, the fascinating entrance to the Atlantic. It was my husband's favourite area in the sea for carrying out his experiments. There were always strong currents there and Guglielmo and I enjoyed the sight of the numerous dolphins jumping high out of the water as they played around the bows of the *Elettra.*

Sometimes we would anchor at Tangiers where we often stayed as guests of the Sultan Menhebi. Then we would set sail again. Guglielmo had a passion for the Atlantic. As he was an excellent captain he could foresee danger in time and he never took unnecessary risks. He had a surprising ability to feel and sense the atmosphere around him. All this made me feel safe as I knew he would take care of me while as for Guglielmo he was proud of my courage.

During these days of feverish sailing the time went by quickly. Many friends still ask me: "How did you pass the time?" Well the time just flew! During the days at sea it was interesting to see, in the chart house and on the charts, the direction in which we were sailing and our exact position. In the open sea we searched the horizon through the binoculars. It was always a pleasure to meet ships both large and small on our course, especially in the Atlantic. We would wish one another "Bon Voyage" via the radio or using flags. It was always a pleasant surprise for the sailors to sight the *Elettra* with Guglielmo Marconi on board.

Nearing the coast, we never tired of admiring the different scenery and we watched the manoeuvres for anchoring the yacht with great interest. My husband, as usual, never allowed the smallest detail to escape him.

When we arrived in port the local authorities came to welcome us at once, putting themselves at our disposal and inviting us to their houses. We were happy to welcome them on board and we always tried to create a pleasant atmosphere.

When the sea was not too stormy we enjoyed long conversations. Guglielmo often consulted the Encyclopaedia Britannica, which he kept in the library on board, because he considered it very authoritative and accurate. We enjoyed reading it, especially before arriving at a port which was unknown to us. Guglielmo, with his youthful enthusiasm, wanted to know all about the history and the customs of the countries we were going to visit.

I can testify, having lived at his side for a long time while he was working, that his technical gifts had something supernatural about them and were accompanied by incredible intuition.

With regard to communications over enormous distances, I should like to record an occasion when I was in Genova in October 1965 during a day dedicated to Guglielmo Marconi and Christopher Columbus. The Mayor of Genova introduced me, with my daughter Elettra, to the American astronaut, John Glenn, saying: "This is the Marchesa Marconi, the widow of Guglielmo Marconi". Mr. Glenn shook my hand and looking me straight in the eye, exclaimed with sincere feeling: "How much we owe him!"

A few years later in Rome, once again with Elettra, I met another famous American astronaut, Neil Armstrong, who said to me: "If it had not been for your husband, I could not have landed on the moon."

# THE ROME EXPRESS AND THE HONORARY CITIZENSHIP OF ROME

Guglielmo used to leave the *Elettra* laid up for the winter in the Port of Genova and sometimes at La Spezia from October to March.

Although my husband had flown with the American, Wright--one of the pioneers of flying--we never travelled by air as in the 1930's flying was still not safe enough.

When we were not on board the *Elettra* we travelled on the transatlantic liners, by car and above all by train. It is impossible to count the number of times we went to England in the "Rome Express".The journey took thirty-two hours altogether. We did it so often that the waiters in the restaurant car on the train knew without being told what our favourite menu was.We were considered honoured guests and all the staff treated us with great consideration. The same compartment, number seven, which was furthest from the wheels, was always reserved for us. We knew the scenery of the journey through France practically by heart.

We left the train at Calais and crossed the Channel on a ferry which we came to know very well. I think it was called the "Maid of Kent". If the visibility was good when we approached the English coast the Captain would invite us onto the bridge to admire the white cliffs of Dover.

Once we disembarked we would catch the train called the "Golden Arrow" as far as London. It was always a great pleasure to get out at Victoria Station.

I have happy memories of the years I spent with Guglielmo in London. Both my husband and I were very fond of the English Capital which was interesting for many different reasons, not the least of which was that feeling of "steadiness", that is, of reliability and safety which in my opinion only London can give. We spent months on end in the English mists. We always had Elettra with us.

My parents often used to take the train from Rome and come to spend some time with us as our guests at the Savoy Hotel. My mother was always full of initiative, and busy making plans for spending entertaining mornings with Elettra. Guglielmo was happy for her to take Elettra for walks in the magnificent London parks around the shady lakes and along the Serpentine, that delightful canal which runs through Hyde Park; or to the Zoological Gardens where, with our daughter and other children, she sat on an elephant and went for a ride through the park. When my husband managed to finish his work in the late morning he was happy to join them with me.

Guglielmo's brother Alfonso also often came to visit us in London. He was always very affectionate towards us and brought lots of little presents for Elettra.

The meetings in London were always a great joy to us because all the family was reunited, especially during the Christmas holidays.

The long winters in England were interrupted by brief visits to Rome, due to the many commitments that recalled Marconi to the Italian Capital, since he was a Senator of the Realm, President of the Royal Academy of Italy and Founder President of the National Research Council.

Many people pointed out to me that Guglielmo visited Italy more often after our marriage. In fact, he was happy, after the long time spent abroad, to come back and enjoy the sun and the beauties of his native land which he had always loved.

When we were in Rome we were my parents' guests at number 11, Via Condotti. We always had the same room with windows overlooking the Oleander courtyard.

Once, on a train, when we were getting close to the Capital, Guglielmo told me about something that happened when he was young, in 1903. He was famous and had already sent the first radio signal across the Atlantic and completed the experimental voyage on board the "Carlo Alberto" in the North of Russia.

At that time the Mayor of Rome was Prince Prospero Colonna, who had taken a great interest in Guglielmo's experiments right from the very beginning and had become a close friend of his. He was full of admiration for the young genius and decided to confer the Honorary Citizenship of Rome on him.

The ceremony took place with great pomp on 7th May, 1903 in the Capitol, in the Hall of the Orazi and Curiazi and in the presence of His Majesty King Vittorio Emanuele III and Queen Elena.

Guglielmo, for the first time in Italy, gave a lecture on his discoveries. When he finished speaking the whole noble assembly exploded in a loud, interminable applause. Even the Sovereigns clapped.

At that point in the story, my husband gave a slight smile, almost as if he was laughing at himself and what happened next. "When the ceremony was over", he told me, "Prince Colonna and I accompanied the Sovereigns out of the Capitol to their carriage. Then I got into a "landau"--an open carriage of that time. I was in morning dress and my mother sat beside me wearing an elegant hat with a veil. An enthusiastic crowd had gathered outside the palace. At a certain moment some students from the nearby La Sapienza University unhitched the horses from our carriage and, taking their place, pulled us in triumph for a good way".

I am happy to mention the many other Italian and foreign cities, apart from Rome, which conferred Honorary Citizenship on Guglielmo; among these: Milan, Florence, Livorno, Pisa, Genova, Bari, Civitavecchia, San Francisco and Rio de Janeiro.

# MY PARENTS

There is so much I would like to say about my wonderful, unforgettable parents: My father and mother were exceptional people.

With their great intuition they immediately appreciated Guglielmo's outstanding qualities and they admired his spirituality and morality as well as the confidence he was able to inspire.

They realized how deeply we loved each other so they gave their blessing to our marriage although they knew that by marrying him I would have to change my life completely and face a busy, international and almost chaotic existence. With their consent and the official approval of Cardinal Eugenio Pacelli, later Pope Pius XII, who was the true councillor of my family, Guglielmo and I were married.

When we were in Rome, my husband and I often stayed with my parents in Via Condotti. We found the affectionate and serene atmosphere incredibly relaxing after so many voyages and adventures at sea. Guglielmo was happy there and we enjoyed the friendly family conversation and chatted about everything that happened with great intimacy.

My husband admired them very much and valued their opinion. He had immediately established an excellent relationship with them, based on sincerity and mutual trust, two very important factors in life.

My parents made a handsome couple. Both were tall and distinguished-looking. My father was fair while she had chestnut hair. My mother had been very beautiful when she was young and was always elegant, charming and amusing; qualities she kept all her life. Everyone loved her.

As for my father, he was very highly thought of because of his intelligence, common sense and integrity, linked with a sense of family and tradition. He was a very cultured man who enjoyed reading the works of the great Italian and French authors as well as books on history. He also collected rare stamps from all over the world, an interest he shared with Guglielmo. My husband, as he admired the stamps, remembered his journeys and his extraordinary inventions; through radiotelegraphy it was possible to reach any place in the world, just as stamps could do.

Guglielmo got on very well with all my relatives. He enjoyed coming with me to visit my uncle and aunt, Prince and Princess Barberini, in their palace full of works of art and rich in history and tradition where the large Barberini family lived. Uncle Luigi, my mother's brother, was the second son of the Marchese Urbano Sacchetti and Princess Beatrice Orsini. When he married his first cousin, Princess Maria Barberini, who was an only child, he took her family name so that the title would not die out, thanks to a special edict by Pope Urbano VIII Barberini in the XVII Century.

We also often visited my aunt, the Marchesa Teresa Sacchetti, her son Giovanni and his wife Matilde in Via Giulia, in the palace where my mother was born and in whose ancient chapel she was baptised and later married my father.

We saw many other friends during our visits to Rome. I remember the delightful evenings at Palazzo Colonna as guests of Prince Marcantonio and Princess Isabelle Colonna; the splendid balls given by Jane Bourbon del Monte, Princess of San Faustino at Palazzo Barberini for her daughter Virginia who later married Edoardo Agnelli. I cannot forget all the other invitations to the palaces of my Roman friends and relatives: the delightful dinner parties given by our great friend, Prince Ludovico Altieri in his historic palace in Piazza del Gesù and the dinners at the various embassies; the British Embassy, the French Embassy at Palazzo Farnese, the Embassy of the United States of America, the Spanish Embassy, the Irish Embassy and many others. I also remember the happy times I spent with Guglielmo in Prince Ruffo della Scaletta's beautiful villa with his son Ruffo and his sister, dear Princess Nives, together with all the other friends I still continue to see.

Our little Elettra, so happy and lively, was always with us during our voyages. However, in 1933 when Guglielmo and I decided to set out on our long journey in the United States we entrusted our daughter to my parents' care. They took this responsibility very seriously and they certainly contributed to making our life a full and happy one. Guglielmo realized this and I myself still feel heartfelt gratitude towards them.

When we were at Santa Margherita Ligure we used to anchor the yacht quite close to the enchanting villa "La Pagana." My mother and father often came to visit us and it was always a great pleasure to have them with us as our guests on board the "Elettra." They were energetic and loved taking part in the expeditions that Guglielmo organized in the motorboat to enjoy the beautiful Ligurian coastline, especially the lovely bay of San Fruttuoso. Guglielmo, our little daughter and I were really happy when they were with us.

# MUSIC ON BOARD THE "ELETTRA"

In the evenings Guglielmo often used to play the piano to entertain our little daughter. It was a pleasant pastime which he had always enjoyed as a distraction from his scientific work. Later he would ask me to play his favourite music, usually Beethoven, Chopin and Schubert; while he was listening Guglielmo, who knew and loved music, would follow the score with his eyes. He had inherited his love of music from his mother Annie Jameson.

My husband's older brother, Alfonso, to whom Guglielmo was deeply attached, loved music too; in fact he played both the violin and the piano. He was also a talented artist and enjoyed drawing and painting.

I, too, loved music and shared Guglielmo's love for the opera. We used to talk about it together and we often discussed artistic and cultural matters. Few people know about Marconi's passion for music. It made a deep impression on him and as he listened he was filled with emotion.

My husband loved Puccini's masterpieces and in 1908 he was present at the first performance of "Madam Butterfly" in New York. We never missed an opportunity to go and listen to "La Boheme" together. When the tenor began to sing the romantic aria "Che gelida manina" (what a cold little hand) he always took my hand in his and I could feel his own hand shaking. As he listened he looked before him with an expression of profound melancholy. Those beautiful notes of Puccini's, so full of pathos, struck a responsive chord in Guglielmo's sensitive heart. Beautiful music always filled him with emotion.

My husband was a very good friend of Giacomo Puccini's and since the Maestro lived at Torre del Lago he was often a guest on board the "Elettra" at Viareggio. Guglielmo loved his compositions for their intrinsic beauty and admired him greatly. Puccini confided to him that the wide diffusion of his works in the world of opera was also thanks to the administrative skill of the Ricordi brothers who were renowned connoisseurs of good music which they knew how to present to the theatres.

Guglielmo often used to meet other famous musicians: Leoncavallo, Tosti and Mascagni in Italy and New York. The great Caruso was also a good friend of his. He told me that the tenor, before singing at the Metropolitan of New York had begged him to stand with him behind the scenes. Marconi's presence helped him to overcome those moments of great tension before he went on stage.

# CONFERENCES

Gulgliemo Marconi was a truly great genius; right up to the end of his life his brilliant mind was continuously working on new projects and he was convinced that much of his success was due to our love for each other.

As proof of this I remember a moment during the last summer we spent on board the *Elettra* in 1936 when we were anchored in the harbour of Santa Margherita Ligure. Sitting in his armchair on deck in the sunshine Guglielmo said to me happily: "You know Cristina, they are really pleased at the Marconi Company in London. They say that my scientific work has intensified since our marriage and that I am making progress towards more and more important results".

In fact, right in this period he reached exceptionally interesting and practically unthinkable goals. Among other important discoveries, Guglielmo invented "blind navigation", later known as radar, created the parabolic antenna, i.e. the satellite and started work on a system to extract gold from seawater.

During his experiments my husband, since he was sure of the results, used to work with enjoyment, making it all seem simple and natural.

Guglielmo was reserved as far as his scientific work was concerned. He never spoke to outsiders about the experiments he was working on; he used to announce his new goals in a conference which he held every two years at the Royal Institution of Great Britain in London. He often spoke to me about the enormous success of the famous conference he gave at the Royal Institution on 3rd March, 1903. It was at this conference that he put forward his ideas on the future development of wireless-telegraphy.

He also held conferences in Rome at the Capitol and the Royal Accademy of Italy; in the United States of America at the Columbia University of New York and at various other universities; in Washington at the seat of Congress; in Chicago on Marconi Day during the Great World Exhibition and finally in California when we travelled round the world in 1933.

Guglielmo prepared his speeches for these conferences personally. He sat at his desk for hours in complete silence writing everything down on sheets of white paper in his beautiful hand-writing. He hardly ever made any corrections. There is documentary evidence of all this because the texts are carefully conserved in the archives of the most important scientific centres. My husband in fact wanted to leave written proof of his scientific work.

He used to speak for over an hour, without interruption, in a calm tone. His great personality, self-confidence and calm speech made an enormous impact as he described the progress of his inventions to sovereigns, heads of state and scientific political personalities in Italy and abroad.

Many foreign scientists, including Nobel Prize winners, admired Guglielmo for his determination and ability as a researcher and inventor. When he had finished speaking there would be an explosion of enthusiastic applause and those great men of science would rush up to him to shake his hand and congratulate him.

# GUGLIELMO MARCONI
# HIS WORK CONTINUES

On board the *Elettra* Guglielmo gave precise orders to his English technical assistant, Mr. Isted, who carried them out carefully. Then he himself checked and inspected everything personally; he was kind but uncompromising. This was another of the reasons that the people who worked for him admired and respected him so much.

I can still see Guglielmo in the dining room on board the *Elettra*. Sitting opposite me at the other end of the long rectangular table he is looking through the open door that leads into the radio cabin. He is smiling and his eyes are bright. His expression is intent and he seems to be thinking. Some new idea is certainly going through his mind. I feel I should not interrupt the thread of his thoughts and keep quiet. After a short time Guglielmo, in silence and still with a smile, using the point of a silver fork on the starched Irish linen tablecloth, carefully traces some technical drawings whose meaning only he can understand. He is happy and satisfied. He has begun a new project which remains a secret because nothing is written down in black and white on a sheet of paper; it all disappears when the tablecloth goes to the wash.

Nobody apart from myself could see that something important had happened. This was the reason I always wanted beautiful starched white tablecloths to be used on board the *Elettra;* they were my own personal contribution to Guglielmo Marconi's scientific and creative work.

Whenever possible we went out into the open air on the beautiful deck of the *Elettra*. There were awnings to shelter us from the sun and wind and comfortable wicker armchairs to sit in. Guglielmo read the barometer and studied the atmospheric presure and the winds; they were details which were very important for his experiments. He used to enjoy giving instructions to the devoted boatswain, Patrono, an old sea dog from

Portovenere. He would tell him how to regulate the ventilation of the yacht and how to create shade on deck with the awnings so as to make our life on board pleasant and comfortable.

In every moment of his life Guglielmo's sensitivity, calm and great practical ingenuity enabled him to resolve even little everyday problems.

# THE "CARLO ALBERTO" AND THE MAGNETIC DETECTOR

In July, 1902 the King of Italy, His Majesty King Vittorio Emanuele III invited Marconi to sail to Russia aboard the royal ship, the "Carlo Alberto" to demonstrate long-distance wireless telegraphy. Guglielmo accepted with pleasure and was given officer's rank. Later in the same year the "Carlo Alberto" was placed at Marconi's disposal for long-distance trials in the Atlantic.

King Victtorio Emanuele III had arrived in British waters aboard the "Carlo Alberto" to attend the coronation of King Edward VII in June, 1902. However the coronation had to be postponed due to King Edward's sudden illness and a ceremonial visit by King Vittorio Emanuele III to Tsar Nicholas II was arranged.

When they arrived at Kronstadt, the most important Russian naval base, the "Carlo Alberto" was anchored beside the imperial yacht, the "Alexander". Tsar Nicholas II and King Vittorio Emanuele III who was the Tsar's guest went on board the royal ship. The Russian National Anthem and the Italian Royal March sounded. The two sovereigns visited the wireless cabin and listened to Guglielmo's explanations. He made a special transmission for them, naturally by wireless, receiving signals from the radio station of Poldhu in Cornwall one thousand six hundred miles away. They were both amazed and congratulated him warmly. Tsar Nicholas II bestowed one of the most important Russian decorations on him: the Grand Cross of the Order of St. Anne.

Later, while they were still at Kronstadt, the Russian scientist Alexander Popov came on a visit to the "Carlo Alberto." My husband told me with pleasure that he met him on the deck at the top of the ship's ladder and Popov immediately shook his hand and spoke the following words which I consider historic and definitive: "I acknowledge you as the inventor of wireless telegraphy".

Guglielmo used to smile ironically when he commented on the conjectures of the other scientists on the subject of his research. In fact he had no rivals. He knew that he was not only the first but the only inventor of radio telegraphy in the world. All the other so-called inventors had only carried out studies, attempts and research which might have been interesting but never came to any concrete conclusion.

In the spring Guglielmo had invented an apparatus which meant a great deal to him and was of the greatest importance for the continuation of his work. This was the magnetic detector, which became known as the "Maggie" to all sea-going wireless operators. It was a completely new and original detector which was more perfect and sensitive than the filings tube. It enormously increased the reception potential of the radio stations in long-distance transmissions and was to be of fundamental importance for the success of his other experiments.

While he was on board the "Carlo Alberto" Guglielmo personally built a magnetic detector. He told me with a smile: "I was sitting on deck with Admiral Mirabello. The weather was fine and the sea was calm. Suddenly, I said to him, 'Give me an empty cigar box...' Using pieces that he chose himself from his little laboratory on board, in a very short time his magic fingers had made the apparatus.

Guglielmo often spoke to me with satisfaction and a touch of humour about the time he made a magnetic detector in a cigar box. Admiral Mirabello, who witnessed it, used to tell the story of this exploit of Marconi's with admiration and enthusiasm and he later had the following inscription engraved on the door of the radio-telegraphic station aboard the "Carlo Alberto": "Today, 26$^{th}$ June 1902, Guglielmo Marconi honoured with his presence this royal ship anchored off Poole, inaugurating the first model of the new magnetic detector invented by himself and donated to the "Carlo Alberto" which, the first ship in the world, verified its functioning at sea". This scientific jewel is now kept in the

Guglielmo Marconi Museum at the Ministry of Posts and Telecommunications in the EUR district of Rome.

The "Carlo Alberto" entered the Mediterranean and sailed for Italy. Before Guglielmo disembarked at La Spezia the ship anchored off Forte dei Marmi in front of Villa Morin (later Villa Agnelli) so that he could land with a rowing boat on the beach which was deserted in those days. The sky was blue and the sea was beautiful that day. He went to visit Admiral Costantino Morin, the Minister for the Navy, to thank him personally for putting the royal ship, the "Carlo Alberto" at his disposal so that he could carry out his trials at sea. My husband, in fact, even though he was so busy with his experiments, always maintained strong links with the Italian Navy which had believed in him and supported him from the start. By 1936 he held the rank of Rear Admiral. He told me that in July, 1897, he had carried out a series of transmissions from the naval base of San Bartolomeo near La Spezia to the receiving station on board the ship the "San Martino" of the Royal Italian Navy. The distance was sixteen kilometres and the signals arrived clearly both above deck and within the battleship. There are several photographsof Guglielmo commemorating the event.

On 10th September, 1902 Guglielmo successfully carried out the first wireless telegraphic transmission between Italy and England. He told me he was at Poldhu while the "Carlo Alberto" was sailing off the island of Gorgona. Guglielmo sent the first telegram in history from his radio station at Poldhu to the ship, addressed to Admiral Morin, in which he expressed his gratitude for the interest he had shown in the progress of his work. Immediately after this, to commemorate the event, Admiral Morin in his capacity of Mayor of Forte dei Marmi decided to call the main square, the one near the sea, after Guglielmo Marconi. It was the first square to be named after him, a young man of only twenty-eight!

In 1903 my husband gave further proof of the importance of his invention. He embarked on the "Lucania" which was sailing between England and the United States and during the voyage he organized a radio service from Poldhu and Table Head. Guglielmo spent the night listening to the news that reached the ship from the two continents and every morning his Atlantic newspaper was distributed to all the passengers of the "Lucania." It was the first news review to be printed at sea.

# NORTH AFRICA GIBRALTAR and VIAREGGIO

In the spring of 1929, Guglielmo and I set off on a long and pleasant voyage on board our yacht "Elettra." We embarked in Naples and headed for Tripoli which, at that time, belonged to Italy. We spent the first few days anchored offshore along the beautiful coastline and took the opportunity to visit the archeological sites of Leptis Magna and Carthage which had once belonged to Hannibal, the protagonist of the second Punic War. We also visited the Italian concessions which had resulted from the labours of our really admirable emigrants.

During this period, we were the guests of the Prince of Karamanli and attended a lunch given in our honour by the Governor in his palace and were also invited to the beautiful residence of the young married couple, their Royal Highnesses, the Duke of Aosta and his wife Anna. The Duke's mother, H.R.H. Princess Helene of France, was also present. We were much impressed by their charm and by the warm welcome they gave to us. Guglielmo and I returned their hospitality and were delighted to have them as our guests on board the *Elettra.*

From Tripoli, we sailed along the African coastline and anchored in Tunis which still had its original suk. We visited Sidi Busaid, the elegant villa belonging to Baron d'Erlanger and its surroundings. Then off to Algeria and Morocco. In Tangiers we were guests in the splendid palace of Prince Men Hebi who invited us to a meal which we ate sitting on the floor according to the local custom.

We also watched a military parade of Moroccan soldiers in their spectacular multi-colour uniforms and characteristic fez and riding white Arab horses. From there we went on to Fez and Marrakesh.

As always, we stopped over in Gibralter, an attractive place that we knew and liked since, in the course of the year, we frequently crossed the Straits that were always the starting point of the long voyages so necessary and important for Guglielmo's experiments in the Atlantic. The English residents were used to seeing us anchor our yacht near the fortress which, at that time, was an important strategic base in the Mediterranean. The Royal Navy always offered us a good anchorage and we would stay for a couple of days, fascinated by the characteristic narrow streets along the seafront at the foot of the Rock, with their elegant little shops where it was possible to shop duty-free.

We saw and entertained many friends, including the English Admiral who commanded the naval base. He always came on board as soon as we arrived and we in turn followed our custom of visiting his beautiful residence with its hanging garden overlooking the sea.

How many times after sailing along the coast of North Africa, we then set course for England, hugging the coast of Portugal, Spain, the stormy Gulf of Guascoigne which took three days and nights to cross, and finally reached Southampton on the south coast. We touched Gibraltar again in late September, following the same route, when *Elettra* had to go into dock.

After Gibraltar, we followed the southern Spanish coast and, after having crossed the Gulf of Lion, arrived at Viareggio in Italy anchoring in front of the Hotels Astor and Royal. This was the start of a resting period for us and I was overjoyed to see my parents again after those days on board which were so fatiguing but indispensable for Gugliemo's important work.

While anchored off the small seaside town of Forte dei Marmi, we invited some of our friends aboard the *Elettra* including the Duke and Duchess Ravaschieri, Myriam Potenziani, Prince Rodolfo del Drago and his sister Maria Cristina. After dinner we danced on the bridge to the music of the Savoy Hotel orchestra, transmitted directly from London by Guglielmo's radio station thanks to a link-up that was absolutely unique for those times.

Virginia and Edoardo Agnelli also came on board and we returned their visit going to lunch in the garden of Villa Agnelli at Forte dei Marmi where we met Virginia's mother, the amusingly eccentric Princess Jane di San Faustino and the Agnelli children, Clara, Gianni, Suni, Maria Sole and Cristiana.

# PORTO SANTO STEFANO AND MEMORIES OF FISHERMEN IN THE BAY OF BISCAY

With my daughter Elettra, I sometimes return to the heart of the Maremma, the part of Tuscany which was one of our favourites. We remember the wonderful days we spent there with Guglielmo on board our yacht. On occasion, Guglielmo had to be within easy reach of Rome because of sittings in the Senate or his duties as President of Italy's Royal Academy.

In the springtime, when we usually lived on board and in order not to neglect his experiments in his special floating laboratory, he chose to anchor off the coast of Porto Santo Stefano, the locality he liked best in the Maremma which, in those days, was still wild and unspoiled. We would anchor well out in the bay so as to get a good view of the deep blue sea. In his motorboat, Guglielmo would often take a trip along the coast with its many inlets where the remains of Etruscan towers, mysterious and abandoned, could still be seen.

Usually Guglielmo preferred to avoid the noise and confusion of the harbours but he was always friendly with the fishermen and sailors who greeted him with respect, admiration and gratitude. Who knows how many times some of them had been in danger on the high seas and had been saved thanks to the invention of wireless.

The same thing happened when we were sailing towards England through the Bay of Biscay. Sometimes we met French fishermen taking a risk by venturing so far out in their boats since sailing in this zone is extremely dangerous due to frequent storms and high seas. Guglielmo admired the bravery of these fishermen knowing how hard their life was and he always slowed our engines when we met them. Weather permitting, the trawlers would come alongside and the fishermen would generously offer us fresh fish. In return my husband would give them good wine.

On the bow of the *Elettra*, he had installed a loud speaker which relayed light music transmitted by radio from England. This amused Guglielmo who was happy to offer a little light entertainment to those hard-working fishermen who, recognizing our yacht from a distance, would sail close in order to pay their respects to Marconi in person.

33. Contessa Anna Bezzi Scali with her granddaughter Elettra--11 Via Condotti, Rome, 1932. (Photo Eva Barrett.)

34. In Spain, on the Guadalquivir, near Seville, Marconi with his wife Maria Cristina aboard their yacht on which he improved on wireless radio transmission using the "beam system"--1928.

35. Marchesa Maria Cristina wearing a typical Andalusian shawl (photo Petri).

36. Marchesa Maria Cristina Marconi.

37. Dressed for the occasion, *Il Duce* Benito Mussolini on a visit with Marconi aboard the *Elettra*, 1930.

38. Marconi and wife Maria Cristina, together with Brazilian Ambassador looking on while Marconi throws the switch in his family residence (11 Via Condotti, Rome, Italy), lighting the Statue of Christ, atop the Corcovado in Rio de Janeiro--Brazil, 1932.

39. Marconi with Maria Cristina in front of the Parthenon, Athens, Greece, September 1932, with Greek dignitaries. Far left is famous archaeologist, Sir Arthur John Evans.

40. Marconi, his wife Maria Cristina, and the crew of the *Elettra*. To the left of the Marchesa is Captain Gerolamo Stagnaro; to the right of Marconi is Chief Engineer Giuseppe Vigo; in the back Officers Giovanni Vertutto, Stefano Ferrando, and Alberto Ricchiardi (radio opeator).

41. Marconi and Maria Cristina on the renowned ocean liner, *Conte di Savoia*, en route to New York in September 1933; unbeknown to them, they were also en route to their first voyage around the world.

42. Marconi with his wife Maria Cristina aboard *Conte di Savoia*.

43. Marconi with maria Cristina atop the Grand Canyon in Colorado, 1933.

44. Marconi and Maria Cristina in Hollywood, 1933, at the home of Mary Pickford (to the right), and Paulette Goddard and Charlie Chaplin (from the left). The famous actors autographed this photo for Marconi.

45. Marconi and Maria Cristina immediately after having been driven through the opening of the giant Sequoia in Yosemite, California.

# SPAIN

When Guglielmo's experiments required us to take a trip in *Elettra* through the Mediterranean, we often sailed close to the Spanish coast and greatly enjoyed anchoring in the harbours at Barcelona, Valencia, Almeria, Malaga, Cadiz and Seville. At Trafalgar my husband would reflect on the great naval battles fought there, the heroic deeds and events of the past.

Every year we would sail into the navigable mouth of the Guadalquivir River and anchor off Seville. Here, during the Feria--a springtime feast characteristic of Andalusia--we enjoyed the climate of that city which had been the capital of Spain for several centuries and especially appreciated its enchanting atmosphere in April, bathed in sunshine, full of flowers and sparkling with life. We would often meet the royal family there and they in turn would visit us on board the *Elettra*. I remember a wonderful evening in the magnificent royal palace el Alcazar and other entertaining evenings at Las Duenas, the splendid residence of the Duke and Duchess of Alba.

The Duke of Alba, Jimmy to his friends, was a tall man, full of charm and a brilliant conversationalist. He and my husband were great friends and very much enjoyed each other's company. We also used to see him and his wife, Maria Rosario, in London and Paris. In Seville we would enjoy watching all the many attractions with them, such as the flamenco dancing by talented gypsy couples who expressed all the Andalusian spirit as they moved to the music of the guitars and castanets and, of course, the colourful and exciting spectacle of the bull fights.

The matador would enter the arena and after having saluted the dignitaries he would turn towards Guglielmo and salute him too. Many beautiful women wearing mantillas and flowers in their hair were present as well as many young people and the citizens of Seville anxiously watching the courageous toreros. The matador, before killing the bull, used to throw his hat called "montera" to Guglielmo as an offering of his noble art.

We were also fascinated by the way the local people celebrated their religious faith and admired the beauty of the processions during holy week, with the crowd excitedly applauding as the very life-like and beautiful "Virgen de la Macarena" statue was carried by.

We enjoyed strolling in the Barrio di Santa Cruz with its narrow white streets and flower-filled balconies. Once we were delighted to meet the Queen of Spain with the two Infantas, Beatriz and Maria Cristina. The sky was always a deep forget-me-not blue.

On the way back to England from Spain we would sometimes stop over in Lisbon anchoring in the River Tago and then continue our voyage touching the Spanish ports of Vigo, La Coruna, Santander and San Sebastian. Guglielmo was interested to meet the local seafaring folk, accustomed as they were to battling against the storms and terrible currents of the Atlantic.

In 1929 we stayed a week in the port of Barcelona at the time of the World Fair enjoying the wonderful water displays and artistic illuminations. After this, Guglielmo and I left the yacht and went on to Madrid where he was interviewed by several members of the press to whom he expressed his enchantment with the beauty of Spain and his appreciation of the World Fair in Barcelona.

A few days later we went to Aranjuez where, in the royal palace on October 21st at 3:30 in the afternoon the first short wave wireless transmission between Spain and Buenos Aires in Argentina took place in the presence of H.M. King Alfonso XIII, the Prime Minister General Primo de Rivera, members of the government and the Diplomatic Corps, local dignitaries and various personalities from the world of industry and finance.

Guglielmo handed the microphone decorated with the Spanish flag and the royal standard to the king inviting him to be the first to speak. Once again, by means of the radio system he had invented, Marconi was able to link together two nations, far distant from each other. I treasure a photograph recording this event.

The next day Guglielmo and I returned to Barcelona where we embarked on the *Elettra* and set sail for Genova. There, as was our custom each year, we intended leaving the yacht in dock for the winter months which we spent in London. We always looked forward however to the following March when we would set sail once more.

# THE VERSILIAN RIVIERA AND VIAREGGIO

I remember a most tranquil sail off the Versilian coast. The sea was very calm and at sundown the last rays of the sun tinged the Apuanian Alps in pink, gold and lilac.

Guglielmo showed me the outline of the mountain called "La Bella Dormiente" (Sleeping Beauty) and he would often tell me that "there, beyond the green of the pinewoods, lies Lake Massacciuccoli where, in his house at Torre del Lago, my dear friend Giacomo Puccini composed some of his magnificent operas: "La Boheme", "Madame Butterfly" and "Turandot." He must certainly have been inspired by the peace and beauty of that enchanting landscape."

Then, sailing off the seaside resort of Forte dei Marmi, I saw the beach where, as a child, I had spent summer holidays with my mother and father and brother Antonio. We would meet our Florentine friends there: the count and countess Ruccellai, Marchese Antinori and his wife, the count and countess Pandolfini, and also Prince and Princess Corsini with whom we still spend pleasant summers together. Elettra and her dear friend Anna Corsini enjoyed going for a swim and sailing in "Sasa", the six-metre sailing boat that has won so many regattas. With them were Anna's many nieces and nephews and Elettra's young son, my beloved grandson, Guglielmo. I also remember the Duke and Duchess Salviati who often came on board the *Elettra*. In turn, we would visit their magnificent estate at Migliarino where we had enjoyable times with our close friends Giacomo, Immacolata, Averardo and Igea sailing in a boat on the river Serchio or pheasant shooting.

My friendship with the Salviatis dates back to my girlhood. My mother, Marchesa Anna Sacchetti, was a friend of their mother, Duchess Maria Salviati who was born Princess Aldobrandini. I remember them together speeding round the estate in a gig with Maria Salviati bravely at the reins.

During one of our short stays in Viareggio, a gale force south-westerly wind was blowing. Guglielmo, who never tired of watching the sea, suggested taking a walk along the pier. We put on our raincoats and set off to defy the elements. From the dock we walked along the pier taking our little girl Elettra with us. The waves breaking against the rocks covered us with spray. We were wet through but happy. Walking against the wind, Guglielmo helped me along with one hand and gripped Elettra's with the other. When the waves crashed over onto the pier he had to pick her up in his arms. Our daughter has nostalgic memories of those wonderful moments spent with her father facing the stormy sea.

## SANTA MARGHERITA LIGURE AND PORTOFINO

Having sailed as far as Santa Margherita Ligure, the *Elettra* dropped anchor a little way off since Guglielmo, as usual, preferred to stay at a certain distance from the port. We often went to the Hotel Miramare, which we especially liked for its location close to the sea.

When we took the motorboat and went on shore, my daughter still remembers how people used to gather round so as to see Guglielmo Marconi in person. First they would gaze at him in silent admiration then they would cheer and applaud.

We would go by motorboat from Santa Margherita Ligure to Portofino where we disembarked in the charming little harbour. My husband, holding Elettra's hand, would pause to chat with the fishermen. He took an interest in their lives thereby earning their respect. Portofino, in those days, belonged to them. Nowadays it has become far too fashionable.

We would wander round the narrow twisting streets and stop in the little piazza to watch the fishermen's wives making lace sitting in the doorways of their homes. We greatly admired their skill in creating these beautiful laces, threading the yarns with their fingers and making the reels tinkle. Passersby would smile at us timidly not wanting to disturb Guglielmo. I would smile back and he would simply nod.

We always waited in the same spot for the motorboat to take us back to the *Elettra* and, ever since, this has been known as "Calata Marconi", Marconi's quay.

# THE PRINCES OF HESSE

Some time ago, I went to the island of Ischia with my daughter Elettra. I had long wanted to see this beautiful island which Guglielmo and I had always neglected, preferring to visit Capri.

We paid a visit to H.R.H. Prince Henry of Hesse, an intelligent and much admired artist who, with great good taste, has built an enchanting house on the rock immersed in the greenery of Forio facing the sea.

One evening at sunset we were on the terrace admiring one of the most beautiful views of the island. The green of the orange trees and vineyards sloped down to the tiny beach of Forio. Looking up towards the hillside, we saw the moon rising.

As we chatted, my thoughts went back to the spring of 1929, when we were on our yacht in Genova docked in the small Duca degli Abruzzi harbour in front of the Italian Yacht Club. Guglielmo and I took great pleasure in inviting the parents of Prince Henry to lunch on board. His father, Prince Philippe of Hesse, was an amusing conversationalist while his mother, born Princess Mafalda of Savoy, was a joyful, thoughtful person who greatly appreciated the special atmosphere on board.

After lunch, my husband took them to his laboratory to listen to the first wireless broadcasts, the first signals from far-off lands. They were most enthusiastic.

Nobody could have imagined in those happy days that Princess Mafalda would have ended her life so tragically in the Nazi concentration camp of Buchenwald. She was a wonderful wife and mother and her death left a great void in the hearts of Italians who loved her dearly. Her son, Prince Henry, told her sad and dramatic story in his book "Il lampadario di cristallo." (The crystal chandelier)

# VISITING THE POET GABRIELE D'ANNUNZIO

I shall never forget Gabriele d'Annunzio's invitation to "Vittoriale", his residence at Gardone on Lake Garda, and the hours we spent with this great poet in September 1934.

After entering the grounds, we were getting out of the car when we noticed that D'Annunzio was waiting for us at the front door on which the words "Silenzio finché parli/Clausura finché s'apra" (Silence until you speak/Seclusion until it opens).

"My dear brother," exclaimed the poet, affectionately embracing my husband who returned this warm greeting. They were both so happy to see each other again. In 1920 they had met on board the *Elettra* at Fiume in Yugoslavia when Guglielmo went there to give moral support to D'Annunzio who, the year before, during the dispute between Italy and Yugoslavia over Fiume, had gathered together a small personal army and seized the town holding it for more than a year in the hope that it might remain part of Italy.

D'Annunzio next turned to me, treating me most kindly, kissing my hand and paying me compliments as if he already knew me even though this was our first meeting. He said how glad he was to see Guglielmo so happy and contented to be with me. At first, in his high-flown style, he called me Madonna Maria Cristina, then Monna Maria Cristina and then more simply Donna Maria Cristina and finally just Maria Cristina.

The poet had us visit the whole of the Vittoriale, showing us the salons and rooms furnished in the most original style and decorated with precious pieces in porcelain and filled with flowers. He described everything in his most characteristic voice using expresssions that sometimes were difficult to understand but always highly cultured. From the rooms on the ground floor, looking mysterious, he led us up a few stairs to his sanctum where there was a table scattered with many sheets of paper filled with his large handwriting.

Happy to have Guglielmo's company, D'Annunzio sat down in the chair where he usually composed his poetry. I was standing next to my husband listening to the verses he was improvising for our benefit with much enthusiasm and spontaneity. Guglielmo, sensitive to the originality of this great poet, smilingly expressed his admiration and thanked him for the privilege of being allowed into his sanctum. Marconi and D'Annunzio, even if so different in character, felt themselves to be close in spirit since they were both dedicated creators, although in different fields: one in science and the other in the arts.

In the dining room we had lunch at a small square table. We were a foursome having been joined by Luisa Baccara, a famous pianist who had abandoned her music to be with D'Annunzio. She was a good-looking brunette with a pale complexion, large dark expressive eyes and a manner that was quietly kind and friendly. The piano on which she played, enchanting the poet, was in one of the salons. D'Annunzio entertained us with his brilliant conversation and Guglielmo answered him with logic and good humour.

We began the meal with fish. "Look, Guglielmo," said D'Annunzio, "I had this fish brought for you from Carnaro." My husband was pleased, remembering the past meeting at Fiume. To end the meal, we were served some delicious oranges and the poet began to praise the scent, colour and exquisite taste of the peeled fruit in a poetic description that lasted at least ten minutes so much so that afterwards Guglielmo and I regretted not having written down those words.

Before we left, D'Annunzio offered us some of his latest verses. It had been a most interesting visit. These two great men were happy in each other's company because they shared the same ideals.

This was the last time we saw D'Annunzio.

# MARCONI AND MUSSOLINI

Another memorable visit on board the *Elettra* was that of Mussolini, at Fiumicino near Rome at the beginning of March 1930. We had spent a long period in Genova where Guglielmo was carrying out the short wave experiments in his radio station on the *Elettra* that culminated in the switching on of hundreds of lights in the Town Hall and World Fair pavilions in Sydney.

We had anchored off Fiumicino where we intended staying two days since Guglielmo had to attend the Senate. I was on board awaiting his return. Since I was soon to give birth, we intended to sail on as far as Civitavecchia.

Suddenly we received an urgent message announcing the imminent arrival of Mussolini. He arrived in a naval launch wearing yachting gear--as my husband always did--and came on board with his customary self-assurance. Guglielmo welcomed him in a dignified manner and immediately took him to the radio station. Mussolini observed everything with great interest and attention, complimenting my husband. He was also very gracious toward me.

It should not be forgotten that my husband who, as an illustrious scientist au fait with world-wide research and well-known on the international scene, had continuous contacts all over the world. He was a deep thinker and possessed an open mind showing understanding of all peoples. With his great experience, Guglielmo never lost sight of the wellbeing of mankind. In this respect, I must make clear certain aspects of the situation in Italy and the attitude of my husband in that delicate period.

Guglielmo was against Nazism. He was worried by the understanding between Mussolini and Hitler and thought it could have serious consequences. Accustomed as he was to thinking logically, he feared and foresaw a probable future war. Since he deeply loved Italy, after careful thought some years later he decided to request a meeting with Mussolini at Palazzo Venezia in Rome. He was received immediately in the Map room. The conversation was very forthright. Guglielmo bravely warned Mussolini against allying with Hitler in a war against England with the probable or, more likely, certain intervention of the United States which would have employed all its war potential and armaments. Italy would certainly have lost. As a physicist, my husband was aware of the very advanced research regarding the use of atomic energy as a terrible war weapon.

"You say these things because your mother was English," said Mussolini, refusing to accept the opinion and advice of Guglielmo Marconi. On his return, my husband who always told me everything, regretfully described this meeting to me adding, "Knowing Mussolini's character, I think he will be irremovable. I'm sure going to war with England will be a terrible drama for Italy and our royal family." I, too, felt sad and worried for our country and for the future of us all.

# GOLFO ARANCI

Some time ago, I went back to the island of Sardinia with my daughter Elettra. We stayed at the Hotel Villa Las Tronas, a property that once belonged to Count Luigi Arborio Mella di Sant'Elia, who was head of the household to H.M. King Vittorio Emanuele III, and was also a friend of Guglielmo.

We visited the most characteristic places in the island, admiring the beauty of Porto Cervo, Liscia di Vacca, Castelsardo, Santa Teresa di Gallura, Cape Testa, Porto Torres, Stintino and Cape Caccia. We felt shivers at the sight of the mountains and the maquis round Orgosolo that have become so sadly notorious.

On a religious feast day, we watched the traditional procession in which good-looking dark-haired Sardinian girls proudly paraded wearing beautiful medieval costumes embroidered in gold and multicoloured threads.

When we reached Golfo Aranci on the ferry boat my thoughts went back to the years 1931-33, when my husband and I entered the same port on board the *Elettra*. Navigating round Sardinia was useful for Guglielmo's short wave radio-telegraphy experiments which consisted in testing transmissions and receiving messages on board the *Elettra* from ever-increasing distances.

Setting out from Civitavecchia, Guglielmo first contacted the radio station he had installed at Monte Cavo at Rocca di Papa near Rome. Then we sailed ever further away from the coast while the station at Monte Cavo continued transmitting with the conversations coming over clearly thanks to Guglielmo's continuous adjustments.

Next he decided to install another transmitter on top of Cape Figari, a mountain close to Golfo Aranci. For quite a while we sailed in the stretch of sea between Civitavecchia and Sardinia and Guglielmo continued testing and retouching his apparatus, eliminating interference until the day when, close to the Golfo

Aranci, the *Elettra's* radio station received loud and clear the transmissions sent from Rocco di Papa, a distance of 270 kilometres. I was so happy to be with Guglielmo in that moment of satisfaction when, once again, he had achieved his aim.

Often during that holiday in Sardinia with my daughter, my thoughts returned to the past and in my mind's eye I saw our elegant white yacht with its green keel riding at anchor on the blue of the then deserted bay. A lifeboat would be dropped overboard and Guglielmo would take the oars while I swam and dived. From the bridge, Commander Stagnaro kept a careful watch through his binoculars to make sure there were no sharks in the vicinity.

Guglielmo and I were alone since there were no dwellings along the coast and we would only meet the occasional fisherman, some of whom used to tell the tale of a solitary old shepherd, living in a mountain hut and watching over his flock on the island of Tavolara, who knew Dante's "Divine Comedy" by heart.

We liked to be alone, feeling the peace around us, isolated from the rest of the world. The days would fly by and we were happy finally to have a little time to talk and exchange ideas. We always had such a lot to say to each other and had so much in common--love of reading, history, music...

Guglielmo reflected on the tale of the solitary shepherd, understanding and envying him. He too would have liked to spend more time alone with me but unfortunately his prestigious public duties did not allow it. I felt the same way, believing that Guglielmo's genius required true solitude. In the vast empty spaces of the seas we sailed over, he found the silence and peace necessary for meditation which, in those moments, became the source of fresh ideas and new inventions.

# H.M. KING VITTORIO EMANUELE III ON BOARD THE *ELETTRA*

In September 1929 we were still anchored off Viareggio when Guglielmo received a message from H.M. King Vittorio Emanuele III who was staying with the royal family at San Rossore. The king, who always took a great interest in Guglielmo's work, wished to pay us a visit on board the *Elettra.* A silent, watchful person, he well knew my husband's worth and had bestowed on him the title of Marquis. Every year he would receive Guglielmo at the Quirinal Palace in Rome and talk to him at length about the progress of his important work, always encouraging Guglielmo to continue with his inventions.

That day all of us on board got up early. The King with his retinue arrived by car at the quay in Viareggio at 8 o'clock in the morning and Guglielmo went to pick him up in the "Thorni-croft", our open motorboat. I welcomed His Majesty on board at the top of the gangplank. Strangely enough the King was not in uniform but instead wore a grey suit. He gave me a slight smile showing that he was pleased to be our guest.

Guglielmo immediately took him to the wireless room where he carefully described the working of the various large pieces of apparatus and gave brief explanations as he usually did, stressing the progress on short and very short wave transmissions. Then he asked the King if he would like to listen in the earphones to the first signals from far-off South Africa which one could hear only from our yacht. The King agreed and listened with great surprise and interest, asking Guglielmo about the future possibilities of his work.

After the visit to the wireless room, we went to the dining room where a buffet had been prepared but the King, as usual, declined any refreshment. The three of us stood there while

Guglielmo and the King chatted on various subjects. At a certain point in answer to a particular question from the King, Guglielmo had to go and get a book from his study. I was left alone with the King which was something unusual; so I waited for him to speak first. He looked at me kindly and made a few remarks which I have never forgotten. Then Guglielmo returned with the book.

Later, we joined the other guests who had been looked after by the Captain on the top deck of the yacht. The visit was over and Guglielmo re-accompanied the King to the quay at Viareggio in his motorboat.

# TO VENICE AND GREECE ON THE *ELETTRA*

Between 1932 and 1934, Guglielmo and I spent several pleasant periods in Venice. The Admiral of the Upper Adriatic usually provided us with an excellent anchorage in front of the Customs House, close to the Church of Santa Maria della Salute.

A quick trip in a motor launch took us to Piazza San Marco. Marconi's slim, distinguished figure was well known to the Venetians and they welcomed him with warm admiration as we sat at the Cafe Florian.

At that time H.R.H. Ferdinando di Savoia, Duke of Genova, a close friend of Guglielmo's, was in command of the Upper Adriatic base. He too was passionately fond of the sea and was considered to be one of the most able admirals in the Italian Navy. He often came on board the *Elettra* where he would spend whole days with my husband. Guglielmo would lend him his cabin for an after-lunch nap and then in the afternoon they would continue their conversation on the bridge. The Duke was a fascinating conversationalist and the two of them would chat freely for hours, talking about everything and everybody with remarkable good sense, something that is not always easy to find today.

In those years, among the most admired personalities of the Venetian aristocracy was the fascinating, green-eyed Countess Annina Morosini and the magnate Count Giuseppe Volpi di Misurata whom their friends jokingly called the Doge and the Dogess of Venice. Later on, another truly great Venetian figure was Count Vittorio Cini, who is remembered for his generosity and for the celebrated foundation that bears the name of his son Giorgio.

I remember that during one of our stays in Venice, we saw their Royal Highnesses the newly married Prince and Princess of

Piedmont, heirs to the Italian throne, who were officially welcomed to this city of the doges with great pomp and circumstance. We saw the handsome young couple seated in the 18th century dogal gondola gliding past the magnificent Venetian palaces along the Grand Canal which, in those days, was silent without the motorboats which have destroyed something of the fascination of this lagoon city.

I saw the young royals again at the gala evening at the La Fenice theatre. They stood smiling in the royal box acclaimed by the elegant audience. For the occasion, the Venetian ladies wore their most beautiful gowns and most precious jewels. Goldoni's magnificent theatre, a masterpiece of the 18th century, was a splendid sight with its brightly lit chandeliers.

We were invited to many dinners held in honour of the Prince and Princess of Piedmont, such as that of the already-mentioned Countess Annina Morosini and Count Giuseppe Volpi di Misurata in their sumptuous palaces on the Grand Canal, and also that given by our great friends the Count and Countess Marcello. For these occasions, ostentation was everything. The ladies wore tiaras and their most beautiful jewels while their bemedalled escorts were in tails. Naturally Guglielmo wore the many important decorations he had received, including those from foreign countries, among which the Grand Cross of the Order of Alfonso XIII of Spain and the Grand Cross of the Order of His Holiness Pope Pius XI. I wore a beautiful evening gown designed by Madame Anna di Ventura and the magnificent emerald necklace which my husband had given me when Elettra was born.

To return this hospitality, Guglielmo and I also held a reception on board the *Elettra* to which we invited the Prince and Princess of Piedmont, the Duke of Genova and Venetian dignitaries and friends. The evening was a success and everybody enjoyed themselves particularly because Guglielmo took some of our guests to the wireless room so that they could listen to a transmission from a faraway land. In those days this was considered a privilege since Guglielmo Marconi's floating laboratory on board the *Elettra* was the only one in existence.

I remember seeing the smile of surprise on the intelligent face of Princess Maria Jose' of Belgium as she sat in the wireless room listening carefully through the earphones. Her husband, the Prince of Piedmont, who was later to become King Umberto II, showed great interest in Guglielmo's explanations. Before leaving the yacht, the princess took our daughter Elettra on her knee treating her with great tenderness. Her own children were not yet born.

In 1934, during our final stay in Venice, Guglielmo held an important and interesting conference on his latest micro-wave research. The Doges room in the Ducal Palace was packed with people from the worlds of science, arts, politics and the Venetian aristocracy, all of whom applauded Guglielmo with enthusiasm and admiration.

In mid-September 1932, we left Venice and set sail on the *Elettra* heading for Athens. We left our daughter with her nanny in the care of my parents who were staying in a hotel on the Grand Canal in front of the Church of Santa Maria della Salute. As we left our anchorage we waved to our little girl from the stern of the yacht and I still seem to see her in my mother's arms as she waved back to us. It was always heart-breaking to have to leave her behind.

The beauty of the Serenissima as seen at sunset from the open sea is impossible to describe. I remember it shimmering as if through a pink cloud from which emerged only the dome and bell-tower of St. Mark's.

Our sail along the Adriatic was without incident, the sea was calm and the sky cloudless. Passing through the strait of Corinth was most exciting. Then arriving at the Piraeus, we were able to admire Athens and the Acropolis as the sun went down. Immediately after anchoring, my husband took me to the Parthenon. He wanted to enjoy with me the view from this famous monument at sunset, in that magic instant in which those marvellous stones are still warm and lit by the sun. We knew that September was the best month to visit the Hellenic peninsula. When we arrived at the top of the Acropolis, Guglielmo stood still and silent for some time, moved by the greatness of ancient Greece.

In those years, tourism on a large scale did not exist. Travellers were rare and very few foreigners went to that part which was only reachable by sea. One saw mostly just cargo ships from the Middle East at the Piraeus and the *Elettra* was the only yacht. So Guglielmo and I could visit the Acropolis and the archeological museum in Athens in perfect solitude accompanied by the director.

We sailed round the island of Crete and reached Canea where we had the pleasure of meeting the famous archeologist Sir Arthur John Evans, discoverer of the palace of Minos at Cnossus, thus laying the basis for all our knowledge of the Minoan civilization. We were fortunate enough to have him accompany us during the visit to the recent excavations. He showed us the beauty of the Palace of Minos, where the majestic pillars truly amazed us.

I still feel the fascination of the mysteries of Eleusi, the ancient monastery of Daphne near Athens where there are some marvellous Byzantine frescoes.

On board the *Elettra*, we followed the coast along the Greek archipelago. It was hot and the weather was fine.

While Guglielmo continued with his studies and experiments, we visited other islands in the Aegean. At Rhodes, which belonged to Italy at that time and was formerly the headquarters of the Knights of the Order of Malta, we were the official guests of the governor. Although we slept on board the *Elettra*, we took part in official receptions in the Governor Lago's palace and at the Hotel delle Rose. We visited the hilly parts of the island. Rhodes was enchanting at that time, it was fashionable to stay there and it was attractive to scholars and artists, a real garden full of flowers.

From the port of Lindo, one beautiful calm evening illuminated by the moon, we raised anchor. All around us, there was deep silence. Guglielmo and I were sorry to leave that haven of peace and beauty. But the green island of Corfu awaited us full of attractions. At that time, it was a lonely treeless place. In the centre of the island, immersed in thick vegetation, was the villa of Archelaus, the ancient king of Macedonia who was the ally of Athens in the war of the Peloponnese. He welcomed Greek artists and intellectuals to his court, among them Euripides. His villa was beautiful surrounded by tall trees. In the years preceding the First World War, Emperor Wilhelm II of Germany spent long resting periods there.

Guglielmo and I stayed only briefly at Corfu but continued our trip and soon reached Syracuse in Sicily. From on high, its ancient walls dominated the port which was used only by fishermen with their boats. We stayed there for a few days because its solitude suited Guglielmo for his experiments.

The Prefect came to greet us on board "Elettra" and organized some excursions for us in the city. We were offered the services of the Director of the Museums, a noted archeologist, who accompanied us to admire the ear of Dionysius and the Greek amphitheatre. And he went with us along the river Anapo on whose banks the papyrus grow.

The Museum of Syracuse, though smaller than that in Athens, possesses works of Hellenic art which are perhaps even more rare and interesting. We were surprised to notice how many traces of the ancient Greek domination were still to be seen in Italy.

Having left Syracuse, we sailed round Sicily stopping at the major seaside towns. We anchored in Palermo for three days and were warmly received by the Prince of Trabia, the Prince of Paterno'and by our friend, the beautiful Franca Florio. Then we sailed on to Genova where we left the yacht in dock for the winter.

We were so happy to embrace our daughter again and then all three of us took the train heading for London.

# THE UNITED STATES OF AMERICA

On September 21st, 1933, Guglielmo and I embarked in Genova on the Lloyds liner "Conte di Savoia" heading for New York. Our yacht the *Elettra* was anchored in the small Duca degli Abruzzi harbour and there we had said goodbye to our dear little daughter. She was only three years old and we had entrusted her to the loving care of my parents. I still remember her saying as she hugged us: "Why don't you take me with you to America?"

We left punctually at noon. From the bridge, where Captain Lena had invited us, we watched the crowd that had come to see us off and which cheered enthusiastically from the quayside.

This was the first time we had gone to New York since Elettra's birth. We missed the United States but I was reluctant to go so far away from my daughter and to stay so long without news of her. At that time, air mail across the Atlantic did not exist and a letter sent by sea mail from New York took more than eighteen days to reach Italy. It was possible to communicate only by means of marconigrams or by cable.

Guglielmo too was upset and showed great understanding towards me. To encourage me to leave, he suggested that we should also book our return trip on a ship leaving New York two weeks after our arrival. I happily agreed to this. But the news that we received from my parents was so good that we were encouraged to prolong our stay, accepting invitations to go to Washington, Chicago and California.

Nowadays, only a few hours by air are needed to reach the various continents. But in 1933, one travelled only by train or on board comfortable ocean-going liners such as the "Rex", the "Conte di Savoia", the "Biancamano", the "Roma", all of which belonged to the Sabaudian branch of Lloyds which served North

America. Others like the "Conte Rosso" and the "Victoria" belonging to Lloyds in Trieste linked Italy to China.

The "Conte di Savoia" was a magnificent ship with large salons and every possible comfort. Guglielmo and I had a suite on the deck named "delle Verande". The voyage was pleasant with fine weather and a calm sea. In the mornings, we would go up on deck to play some sport and swim in the pool. After tea in the afternoon, we would watch a film or stroll on deck until it was time to dress for dinner at which we would often sit at the table of good Captain Lena who was always full of kindness towards me. Music and dancing whiled away the evenings.

In 1933, tourism organized on a grand scale did not exist. Few persons made the lengthy trips to Paris, London or New York. But for Guglielmo, this was his eighty-seventh crossing. He would frequently tell me how his many voyages in the Atlantic had always been for work reasons, making me understand how often his life had been hard, dedicated only to research. Now, at last, without losing sight of the scientific scope, he was able to combine his work with the joy and serenity of a pleasant voyage with me.

With us on the trip were my maid Eugenia, Guglielmo's secretary Di Marco, and Lettieri, the detective in charge of our security. We had a great deal of luggage, trunks and suitcases, Guglielmo's small items of equipment, books, papers and a great many clothes. These last were necessary because we had to change frequently for the numerous private invitations and the many official ceremonies.

One of the most exciting moments I remember of that voyage was when we heard the voices of our daughter Elettra and my parents who called us from their home in Via Condotti in Rome. The "Conte di Savoia", was the first ocean-going liner to be fitted with a special cabin adjacent to the wireless room. It gave Guglielmo great satisfaction to see the results of his studies concerning the wireless-telephone applied for use by all passengers. Our long trips in the midst of Atlantic storms, the constant and fatiguing experiments to perfect the beam system he had created, had resulted in this magnificent achievement which we were able to enjoy in that moment.

Guglielmo was pleased about this and often took me to the ship's wireless room to be present when the other passengers made their telephone calls especially those of emigrants who were on their way to join relatives who had been already living in the United States for many years. As soon as they heard the voices of family members, these people would start to cry, yelling "Hello? hello? Can you hear me? I can hear you!" They were so emotional they could say nothing more. So Guglielmo would calmly advise them not to shout since they could clearly hear the voices of their relatives even if they were far away. Then he would turn to me and say, "This equipment is very sensitive and is damaged by shouting into it. It suffers." More than a week had passed when the sight of the Azores told us we were nearing arrival. Welcoming telegrams began to arrive for us from every part of the United States.

On September 28, the "Conte di Savoia" anchored in the port of New York. Thousands were waiting for us on the quayside and gave us an enthusiastic welcome. Members of the press came on board and clustered around Guglielmo, showering him with questions. They wanted to know about his scientific projects and his latest inventions. As usual, he answered with his habitual competence, courtesy and good humour. They were also very nice to me and even wanted me to make a brief speech on the radio.

As soon as we disembarked, the police helped us through the applauding crowds. I shall never forget the Americans who greeted my husband with a show of their esteem and deep gratitude. It was so moving and meaningful to observe that great population honouring Marconi, the benefactor of humanity, the saviour of those who found themselves in danger on sea or land. Great America paid homage to Italian genius! The motorcycle outriders, sirens blaring, escorted our car, stopping the traffic. We crossed the city of New York between two lines of wildly cheering people. The square in front of City Hall was packed with an enthusiastic crowd and the famous Fire Brigade with its big red fire trucks, so important for the security of that immense city, was also present.

Jimmy Walker, the Mayor of New York, made a warmly welcoming speech and other speeches--from the local authorities and Guglielmo's thanks in reply--followed. We spent several days in New York staying at the Ritz. The governor of the State of New York and other personalities invited us to dinners and receptions.

I remember the first time I went to America as a young bride in 1927. At Columbia University, where Guglielmo had often held important conferences on his discoveries in wireless-telegraphy, a solemn ceremony took place when Chancellor Nicholas Murray Butler conferred an honorary degree on my husband in the presence of distinguished members of the scientific world. I felt very proud of him as I watched him receive his degree.

In 1909, Guglielmo was awarded the Nobel Prize for Physics by the King of Sweden. It was really something exceptional that so young a man--Guglielmo was only thirty-five--was already acclaimed as a great genius worthy of this highest recognition in the world of science.

Going back to the trip to the United States, I cannot forget the affectionate welcome of the most distinguished families of New York, such as the Vanderbilts, Morgans and others, all of whom invited us to their luxurious residences. In the charming localities of Long Island, New Jersey and Connecticut we visited friends and acquaintances in their summer residences.

After New York, Guglielmo and I went by train to Washington D.C. where we stayed at the Mayflower Hotel. A meeting had been arranged for us to be officially received by the U.S. President, Franklin Delano Roosevelt. The president received us at the White House in his private study in the large Oval Room. Roosevelt was seated behind his desk in his wheelchair. The American flag with its stars and stripes was on his right. He greeted Guglielmo affably and immediately a reciprocal liking was established between them. It was an unforgettable occasion.

That evening a reception was held in the president's large private apartment where we were welcomed at the door by the First Lady Eleanor Roosevelt, who greeted us most cordially. She was especially affectionate towards me. President Roosevelt was waiting for us in one of the salons surrounded by senators and congressmen. After the presentations and a drink, we crossed various large rooms on our way to the dining room with a robust-looking serviceman pushing the president's wheelchair. I was surprised to find so much energy and vitality in so seriously disabled a person since the president was no longer able to walk.

At table my husband and I were seated in the places of honour and Roosevelt was truly happy to spend that evening with Guglielmo. Eleanor Roosevelt was celebrating her forty-ninth birthday that evening so she was in a happy mood. The guests had been carefully chosen. It was a brilliant occasion and naturally at the end there were numerous toasts extolling Marconi and, with much kindness, the company toasted me as well.

After dinner, we stayed on in that pleasant atmosphere. I chatted at length with Eleanor Roosevelt who expressed great interest in my life with Guglielmo. She appreciated and agreed with my choice to dedicate myself completely to my husband in order to give him ever greater courage since the successful outcome of his latest research was very important for scientific progress.

That same evening, as a demonstration of his esteem for my husband, the president proposed that we should go to California, putting a special train at our disposal. This meant a five-day journey across the United States. I immediately thought of Elettra, so far away. Prolonging our trip meant adding to the anxiety of being even further away from our daughter. Mrs. Roosevelt, seated next to me on a settee, kindly but at the same time firmly and understanding my sacrifice, succeeded in convincing me. With great humanity, she gave me courage, finding persuasive words and smiling at me affectionately. We met various local personalities in those days spent in Washington, including the magnate who, among other activities, had set up the Mellon Institute and the National Gallery.

Naturally, the Italian Ambassador, Augusto Rosso, invited us to the embassy in a beautiful Italian-style building which, some years earlier, the then Italian Ambassador, Duke Gelasio Caetani di Sermoneta, had had built. We were present at a gala dinner together with members of the United States government and the Diplomatic Corps.

Most interesting and moving was the session of Congress at the Capitol where speeches were made praising Marconi and his great work. Guglielmo answered in his perfect English, remembering and emphasising his deep friendship and sincere admiration for the peoples of the great American nation. At the end, everybody rose and enthusiastically applauded. Pleasant things were said about me too. Guglielmo and I were most moved and happy.

That day Congress was packed. All the dignitaries from Washington and other U.S. cities were present. There was profound and sincere admiration for Guglielmo. He, in turn, well knew the qualities and the psychology of that still so youthful population. "They are a great mix of different races," he explained to me "that's why they are healthy and good-looking. That's why they have new ideas, make progress and go ahead of the others."

Although my husband was a genius who looked to the future, nevertheless he also gave great importance to traditions. He loved and admired England without being hostile to modern-day reality and the rapid progress of the United States. Once again this demonstrates how open-minded Guglielmo was and how great was his understanding in judging the problems of each population.

By October 1st we were in Chicago where we spent three very busy days visiting the World Fair, an enormous exhibition on true American scale entitled "A century of progress." It spread over a large part of the city including Lake Michigan and exhibited all the achievements of progress in the twentieth century. Only a country like the United States could put on a similar show. Guglielmo Marconi was a living symbol of progress and it was a real triumph for him.

As soon as we arrived, there was the welcoming ceremony at the Drake Hotel where we were staying. In the afternoon we cruised on Lake Michigan on board the yacht "Mizpah." In the evening, in the course of a banquet, Guglielmo was awarded an honorary degree from Loyola University and the festivities in honour of my husband culminated on October 2nd in a special "Marconi Day." At the official reception we were the guests of honour together with the president of the United States, Franklin D. Roosevelt and his wife Eleanor. The most important personalities in science, politics and culture had come to Chicago from every part of the world. There were toasts and speeches while the orchestra played. To end the day we were present at the great illuminations of Chicago and the World Fair. Thanks to a clever arrangement based on my husband's experiments, the luminous energy of the star Arthur was transformed into electric energy and all the city lights came on at the same moment. Guglielmo Marconi was feted with admiration, gratitude and enthusiasm for all his discoveries and with an incredibly lively dose of human understanding as this great nation of the future knew how to do.

On October 3rd there was a luncheon at the Dante Aligheri Society and then a reception in the Italian pavilion at the Fair. In the evening, during the official dinner, Guglielmo received another honorary degree, this time from the Northwestern University of Chicago. When we got back to the Drake Hotel, we were so happy to find a marconigram from Rome in which my mother assured us that Elettra was well and looking forward to our return. On our return to New York, we were overwhelmed by a fresh whirl of invitations.

The press wished to meet us again in order to hear about my husband's future projects. We received the journalists in our suite at the Ritz and they listened with interest, esteem and veneration as Guglielmo underlined how anxious he was to do research for the progress and well-being of mankind.

One evening, during a reception, one of them asked Guglielmo a strange question. "Will it be possible to retrieve the communications lost in space at a distance of years?" My husband, visibly amused, replied, "Nothing is more probable. Already today we collect echos which are nothing more than lost communications. Nor can it be excluded that in the near future, we might arrive at person-to-person communication, that is to say each person will go around with his own little pocket apparatus on which to fix appointments, give orders to his broker and make dates with his girl friend." The journalist went away satisfied but he certainly could not have imagined that in that moment Guglielmo had anticipated one of the many developments of his invention. Everybody who spoke to him was struck by the clarity of his ideas, his forecasts for the future and the precision with which he explained all this.

The press was also very considerate towards me since it was appreciated that my close union with Guglielmo had helped him both as a man and as a researcher.

My husband had heard a great deal about California and wished to visit this land of the pioneers who had faced the adventurous crossing of the American Continent towards the Far West. He had a great desire to make that journey with me by train. So he reconsidered the proposal that Roosevelt had made us a few days earlier. The president kept his word and put a special train at our disposal containing two large compartments for Guglielmo and me, two comfortable bathrooms, a dining room, lounge, an observation car and even a well-stocked library as well as compartments for my maid, Guglielmo's secretary and our detective. The service was excellent.

That train trip, from East to West, was a triumph and was to be very useful for Guglielmo. It lasted seven days and seven nights. We set off on the evening of October 13 and woke next morning in open country. We had a brief stop at South Bend in Indiana and the weather was good all the way. My husband had been invited to receive another honorary law degree from the renowned Catholic University of Notre-Dame.

Then we began the long crossing of the Arizona desert which was an impressive and atttractive sight. That vast plain changed colour continually, green in some parts, it then became variegated and the sky was tinted deep red during the marvellous sunsets. I can still see Guglielmo sitting in the observation car at the end of the train. Dressed in blue as usual and seated in a comfortable armchair he gazed out in silence at the confineless space. He wanted me too to look at those colours, those sunsets. He would smile at me and I watched him in silence. I felt that in those moments, faced with the beauty of nature, fresh ideas were springing into his mind.

During that long trip by train, we stopped for two days at the Hotel Eltovar situated on the edge of the Grand Canyon. This huge chasm the same colour as the rocks of the Dolomites had been formed over the centuries by the Colorado River. Rightfully enough, it is considered one of the wonders of the world and is absolutely stupendous!

The clear sky and the bright October sun made it possible to drive many kilometres in an open Cadillac in order to enjoy the various views from different points of the Grand Canyon. We watched a sunset of incredible colours that were reflected iridescently on the rocks until the sun went down completely. By moonlight at night one could see the river Colorado deep down in the grand Canyon as it flowed forming a luminous, silver streak. Silence and a surprisng calm reigned. In 1933, there was less travelling than today and there were few noisy tourists.

Guglielmo was happy. I am sure that in those hours of serenity and solitude in the presence of that marvellous sight, with nothing to disturb him just as when he was sailing on board the *Elettra*, he charged up his batteries and accumulated fresh energy that he then used in his experimenting.

The fastest railroad westward went through New Mexico. We stopped at the Frederick Harvey Hotel at Santa Fe and then at Albuquerque where we visited the reservation of the Pueblo Indians. (Many of them had been converted to Catholicism by the missionaries). We went into their villages in order to observe close-at-hand their customs and habits. My husband amused himself talking in English with the chieftains, the mothers and children. When they realized they were talking to Guglielmo Marconi they were amazed and some even kissed his hand at which Guglielmo simply smiled back with the humanity which was typical of him. I still have some small silver and turquoise bracelets made by Red Indian craftsmen and given to me by Guglielmo as souvenirs.

Towards evening we returned to our train which had been parked in a siding in the shade of tall eucalyptus trees at Santa Fe station. During the night when we were sleeping our train was hooked up to a train coming from the East and heading Westward.

Before reaching Los Angeles, the train stopped several times at the stations of small towns along the way. Hundreds of people came to greet us. I was glad to pass through Pasadena also because I hoped to see my uncle Prince Domenico Orsini again who was assistant to the papal throne and who had married a pretty American woman, Laura Rowen, from Los Angeles. Unfortunately, they were away in Europe.

At Pasadena, there were scenes of great enthusiasm. An enormous crowd, above all Italians, was waiting for us with flags and flowers. When we came out on the observation car platform they clapped and cheered, shouting "Hurray!" and "Welcome!" Grateful for that spontaneous tribute, Guglielmo alighted from the train and was immediately surrounded and almost submerged by a multitude of people anxious to see him close to and touch him. He smiled calmly, waving and shaking the hands held out to him and stroking the heads of babies who stared at him wide-eyed. Our compartment was filled with flowers and plants.

In Los Angeles, we found another huge excited crowd including many Italian immigrants especially farmers. The offical welcoming committee was headed by the mayor. People gathered closely around Guglielmo anxious to greet him and they were very kind to me too making warmly admiring remarks.

I shall never forget the beauty of the flowers in those parts. So many flowers! The lounge of our suite at the Ambassador Hotel was like a conservatory. All the local varieties of fruit and flower had been arranged in large baskets.

We were invited by Robert Andrew Millikan, Nobel Prize winner for Physics, to the Californian Institute of Technology near Los Angeles. He had us visit his experimental laboratory. Millikan was a nice person and he and Guglielmo had long been friends and they spent a long time talking together.

At this point, I recall that in the spring of that same year we had visited Cambridge in England where the university conferred an honorary degree on Guglielmo. We were the guests of the Chancellor of Kings College. The ceremony at the Senate House was solemn and beautiful with the highest dignitaries of the university making complimentary speeches--as had happened on similar occasions--full of esteem and profound admiration for

Guglielmo's genius in recognition of the greatness of his work. The universities of Oxford, Glasgow, Aberdeen and Liverpool also had conferred honorary degrees on Guglielmo.
In Cambridge, together with other scientists, Nobel Prize winner Lord Rutherford was present, the author of important studies on the structure and splitting of the atom. He too was a good friend of Guglielmo's and he invited us to visit his laboratory. My husband was enthusiastic as he looked through the microscope handed him by the scientist. I watched them and heard their comments, the ideas they excitedly exchanged as if they were talking about the most ordinary things in the world. It's quite true that people who are really great are always gifted with simplicity and clarity.

But to get back to Los Angeles, before we left, Robert Millikan wanted Guglielmo to visit the famous observatory at Mount Wilson that towered over the city. It was a clear night. The steep and twisting road right on the edge of the precipice was startling. We were in a small bus driven by an expert driver who had made that trip every day for eighteen years and therefore knew the road well. But, it was breathtaking to look down into the void. When we got to the top of the mountain, we were compensated for the fear we had felt. From up there the night-time view of Los Angeles and the Pacific Ocean was beautiful. At that time, the observatory was one of the most important in the world. We gazed at the stars through the telescope, observing the rings of the planet Saturn, the mountains on the moon and other marvels of the firmament. Then we had some refreshment at the shelter. Guglielmo was in a very good mood. We returned to the hotel only at 4 o'clock in the morning.

We visited Hollywood and in one of the huge studios belonging to the RKO film company, a luncheon was held in our honour attended by the most important producers and film stars. Guglielmo was seated next to Mary Pickford and I was next to John Barrymore so I had the chance to take a good look at his characteristic profile. Later on, the famous impresario Samuel Goldwyn had us visit a studio in order to show us filming technique. At that precise moment, John Barrymore who belonged to a great family of actors, was rehearsing a scene. In front of us visitors, the most handsome and most famous of all the Barrymores muffed his lines. The director yelled "Stop! Cut!" The cameraman stopped filming and Barrymore, annoyed, exclaimed "Nuts!" There was a moment of embarrassed silence in the studio. Then the actor regained his composure and came towards me proffering his excuses.

That day I was wearing a light pearl-grey dress and a small matching hat. I noticed that Samuel Goldwyn was looking at me with intense professional interest. Saying goodbye to me, he said that he would have been very happy to use me as an actress in a decent and important part, promising to pay me a very large fee in dollars. Naturally this was an impossible proposal and Guglielmo and I smilingly thanked him. We really enjoyed this episode.

Mary Pickford, the famous blonde actress, was really quite small with a very clear complexion and a certain fascinating grace. She invited us to have tea at Pickfair, her residence in Beverly Hills. All the Hollywood stars had luxurious homes there. Pickfair had a beautiful front garden full of flowers. Mary Pickford, with her enchanting smile, came out to meet us on the front steps. Charlie Chaplin was by her side and there were many cinema celebrities among the guests including Paulette Goddard, then at the beginning of her career and engaged to Charlie Chaplin. The atmosphere immediately became warm and friendly.

Guglielmo talked to Charlie Chaplin for a long time and was struck by the culture and intelligence of that most popular actor, capable of both amusing and moving audiences all over the world. At a certain point, Chaplin took his bowler hat and walking stick and began to walk around the room with the characteristic gait he used in films causing great admiration and merriment.

That same evening, we took the train and left for San Francisco. The railway line ran close to the Pacific coast, passing through fertile countryside cultivated for fruit-growing with orange, tangerine and lemon trees and lots of almond trees in blossom. It was like being in Italy. The Italian immigrant farmers had had the merit of being the first to transform that fertile land, making it resemble their native land of origin.

On October 24, in San Francisco, we received a welcome without precedent. More than ten thousand people came to welcome us. This exuberantly festive crowd clapped and shouted "Long live Marconi!" "Bravo!" For us, it was a surprising display of esteem and affection which we never forgot.

We were the guests of City Hall. The Mayor, Angelo J. Rossi was an American citizen born in San Francisco of Italian parents. He had done a great deal for his beautiful city. In greeting me, he gave me a corsage of orchids that I pinned to my dress. I was wearing a black dress trimmed with leopard fur and a small green hat with a pheasant's feather. We stayed at the Fairmont Hotel situated high up overlooking the famous and beautiful bay. That evening in the salons of the hotel, we took part in a banquet in our honour with more than seven hundred guests.

The next day we were received in great spendour at San Francisco City Hall. The Governor of the State of California, Mayor Rossi and other city fathers were present. The inside of the huge City Hall was packed with thousands of people who wildly applauded us. High-sounding speeches were made in Guglielmo's honour and he replied in his impeccable English.

My husband knew the American mentality well. Using his intuition, he knew how to choose the expressions most likely to move the hearts and minds of the public and he never forgot to say something amusing. So he was always a great success.

American speakers, as did those in other countries we visited, at the end of their speeches always added some kind remarks about me. Guglielmo would thank them on my behalf saying flattering things which moved me.

In the afternoon, an honorary degree from the University of St. Claire and Stanford was conferred on Guglielmo, an event that was naturally accompanied as always with a wonderful reception, excellent speeches and an applauding public.

A two-day trip was organized for us to Yosemite Valley Park so that we could admire the gigantic thousand-year old sequoia trees, the beautiful waterfalls in the forests and the brown bears. A photograph was taken as we drove in a car under the trunk hole of the biggest of those trees. Another interesting excursion took us along the stupendous panoramic Seventeen Mile Drive as far as Carmel where there was an ancient Carmelite convent.

Among the personalities in San Francisco, we saw a lot of Mr. Giannini, a most talented person who had founded the Bank of America, and who directed it with energy getting it back on its feet after the disastrous earthquake of 1908. His daughter Claire was also very intelligent and Giannini's grandaughters are still Elettra's friends, above all Anne Giannini who is married to Jim McWilliams.

The building of the new Golden Gate bridge crossing the bay of San Francisco had just terminated but it had not yet been inaugurated. One evening the mayor took us along this bridge in an open Cadillac so that we could admire the greatness of this new construction and the beauty of the bay in the moonlight.

It was an enchanting evening. Seen from the Golden Gate under the clear night sky, San Francisco looked much more solitary than it does today: silent, beautiful and absolutely fascinating.

# A VOYAGE AROUND THE WORLD: JAPAN

During our stay in San Francisco, the Japanese Foreign Ministry, on behalf of its government in Tokyo, invited us to visit their great country, and an important Japanese benefactor, Baron Okura, offered us his complete hospitality.

I was undecided. To return home via the Far East meant prolonging my separation from our daughter Elettra. She was constantly in my thoughts and I missed her so much. After having thought about it at length, we decided to accept. It was a unique opportunity. To convince me Guglielmo had said, "We will see new countries and new things together and meet interesting people. I would really like to make this trip with you." This was the eighty-seventh time Guglielmo had crossed the Atlantic by sea.

Our love for each other convinced me. This was a very happy period of our lives. Everywhere we were treated with affection by all the great populations we came into contact with and we found this extremely moving. Even so, we often thought of Elettra. What a long time without seeing her!

Before leaving the United States and setting out on the long crossing of the Pacific Ocean, I had to know that my daughter and parents were all well. So using his expertise, Guglielmo organized a really special two-way telephone bridge between Italy and California. It was a great event. He used the beam system, linking the stations in the United States and Great Britain to reach Coltano, near Pisa which, right from the start, had always been the most important in Italy.

The date with Rome was fixed twenty-four hours ahead of time but we did not calculate properly the differences between the time zones. For this reason, we woke my parents at three in the morning in their home in Rome in Via Condotti. My mother assured me that they were all well and she encouraged me to set out on this new journey, saying,"Do come back via Japan, Elettra is fine." With these reassuring words she dispelled all my doubts.

Guglielmo was very happy. Once again, thanks to his brilliant discovery, he had been able to annul the enormous distance that separated us from our loved ones. His only regret was that such an exceptional instrument was not yet available to all because it had brought so much peace of mind to us. The following day the papers wrote about this communication as an exceptional event, praising the genius of Guglielmo Marconi.

Great Britain and the United States had been the first to recognise the importance of Guglielmo's discovery of wireless, admiring him as a "god", a conqueror of the ether. The English particularly appreciated the possibility of putting themselves in direct touch with their colonies in Africa, India, Australia, Canada and all the rest of the British Dominions. It was truly an imcomparable advantage for the whole of humanity. By now there were no more distances between populations. Space had been annulled! Newspapers all over the world published enthusiastic articles.

As a memento of that journey around the world, I have always kept the scrapbooks given to me at our departure from the places we visited. They contained the photographs and newspaper cuttings that mentioned us.

Our departure from San Francisco on 2nd November 1933 was very exciting. We were sorry to leave the many American friends who had been so hospitable and who had made our stay in enchanting California so pleasant. I repeat, very few Europeans made that long sea voyage across the Pacific and few Americans went to Japan. A certain amount of courage was needed to make that journey in those days. One was cut off from Europe for some months. Letters from the Far East took two months to arrive. My husband feared that my maid Eugenia--who was with me for twenty years--would be upset at the idea of going even further away from Italy and would have made difficulties about going with me. I managed to persuade her by calmly explaining that we were setting off for home but that instead of taking the same route we would have returned by a different one, thereby lengthening the journey a little. We had to take advantage of her scarce knowledge of geography and so she willingly came along, trusting in us.

For the voyage across the Pacific, Guglielmo chose the Japanese ship "Chichibu-Marù" (from the nickname of Akihito, the heir apparent, son of Emperor Hirohito) so that we might immediately immerge ourselves in the atmosphere of the Japanese empire.

When the ship left the quay at San Francisco it was very moving. In our honour, scores of coloured streamers were strung from the bridge of the ship to the quayside seeming to form a multi-coloured forest. All this was accompanied by the sound of sirens and by festive shouts of "Bon Voyage! Happy Journey!" My husband and I were leaning side by side over the rail on the poop deck looking at our friends as they grew indistinct as we drew away.

To have the recognition of people was the best reward for Guglielmo for all his efforts, his masterpiece. He felt that this was true gratitude and spontaneous admiration. Leaving the Bay of San Francisco through the Golden Gate was an exciting sight. The sun was sinking in a real autumnal sunset with great black and purple clouds forming a frame to this most beautiful city built on the sea.

When San Francisco was no longer visible we went up on the bridge and joined the Japanese captain who was waiting for us in the chartroom to show us the route he intended following. He knew that Guglielmo Marconi was an experienced sailor and would have appreciated personally checking the navigation. A little later, looking at the Pacific towards the horizon, we noticed a bank of very black clouds in the middle of which the bow of the transatlantic liner seemed to sink. It was really impressive but voyages on the high seas had never frightened me. Quite rightly, Guglielmo, with his great experience, who always told me that during the big storms there is no danger if the ship is far away from the coast provided, of course, that it has a good captain and crew. His real worry was when navigating in fog or near rocks.

The Pacific Ocean is renowned for being quite the opposite of pacific. We were swallowed into the heart of the tremendous storm that followed us for the whole of that first week. The rolling, bucking and butting was unbearable for almost all the passengers who were sick and suffered terribly. Only Guglielmo and I stayed well, accustomed as we were to sailing on the *Elettra.*

The air was vibrant and cold but the inside of the ship was centrally heated. We felt full of energy, running up and down the ships' ladders and trying to keep our balance on the decks. The battering of the sea was very strong with waves the size of huge mountains. We never tired of admiring that immense dark green ocean and the sky with the strong colours of the sunsets so different from the Atlantic Ocean. Guglielmo was finally having a very rare vacation period.

The "Chichibu-Marù" was Japan's most modern liner belonging to the major shipping line Akusan-Marù, plying between Japan and the United States. The captain and crew were all first-rate with absolutely perfect discipline and organization; Guglielmo considered them to be excellent sailors.

On board we had the main suite situated at the centre of the ship. It was extremely comfortable with big cabins containing lounge and bathrooms. The walls and furnishings were in shiny, exotic Japanese wood, different from ours, with a characteristic oriental smell. Panels and furniture were decorated with pagodas and typically Japanese scenes featuring the kind of landscapes we would later have the pleasure of admiring in person.

For the whole of the voyage, the very courteous captain invited us to his table in the main dining room where we would chose between Japanese or French cuisine. Guglielmo and I were entranced and amused by the general change of scene. We knew that if all went well the crossing would have lasted twenty-nine days, and this made us happy. We both loved the sea and life on board. We felt far away from everybody but we loved each other so much.

In my morning prayers, I entrusted my loved ones to God. There was no alternative since we could not even communicate with Italy because none of the instruments Guglielmo would have required were on board. Our fellow passengers came from many nations: Japanese, Hawaiians, Chinese, Malays, Americans, English, Canadians and a few Europeans.

After the great storms, the sea became calm and the other passengers emerged from their cabins. Chatting with some of them, we began to know the habits, customs, ideas and mentality of the various populations of the Far East with whom we would soon come into contact. Guglielmo rightly observed how easy it was to make friends with people who travel a lot, who are interested in visiting new continents, thereby broadening their horizons and enriching their spirit, becoming more human.

As I have already said, in 1933 when we began this round-the-world journey from East to West, one had the benefit of being able to travel only by sea or train. This made it possible therefore to assimilate the characteristics of the different civilizations and get to know the habits. Nowadays, travelling by air, one reaches continents in a matter of hours, meeting up with realities that are unknown to us. Suddenly on landing, one loses the fascination of long journeys.

After almost a week at sea, we put in at the Hawaiian Islands where Guglielmo was amused by something that happened as soon as we arrived. We were still anchored off shore and one of the first persons to arrive on board and who dared to knock on our sitting room door was the manager of the Ritz in Honolulu, a beautiful hotel on the sea front close to a palm grove with all the characteristics indicated by the famous Mr. Ritz. The manager had known Guglielmo for many years since the time when he had served him several times at the Savoy Hotel in London. Very respectfully he enquired whether Guglielmo would do him the honour to dine at the Ritz. He even remembered Guglielmo's favourite menu: la petite marmite, roast chicken with mashed potatoes, dessert and white wine from Moselle. He was convinced that during the long voyage in the Pacific on board a Japanese ship, we would have had nothing to eat but oriental food.

In Honolulu we were the guests of the U.S. Governor, Mr. Judd, who was most attentive towards us. He took us by car to show us the most attractive places and how the local people lived in those enchanting islands. Hawaii was still primitive and wild in 1933. Tourism was non-existent and there were very few white people. We watched the traditional dances of the local people who wore straw skirts and garlands of flowers. Good-looking youths climbed quickly and agilely to the top of very tall coconut trees and it was astonishing to see them surfing on the waves balanced on the boards they had made themselves. We never imagined that surf-boarding would have become a sport practised all over the world.

We listened to the local music with the sound of the melodious guitars and songs sun by the best Hawaiian performers. The luxuriant tropical vegetation with its brilliantly-coloured exotic flowers was truly beautiful.

The first evening Governor Judd invited us to his typically colonial residence belonging to the U.S. government where we attended a reception in our honour at which were present all the foreign residents. All were fascinated by the enchantment of these extraordinary islands.

In Honolulu Guglielmo gave me the pleasant surprise of putting me in contact with my parents and our daughter Elettra. In order to do this he used the beam system as he had done in San Francisco. With this very special, if not unique, link-up I was able to speak with my loved ones in Rome.

I can still see him, watch in hand, waiting near the microphone. Many important stations in the United Staates and in England had already been notified and knew that some days earlier Guglielmo Marconi had organised this special wireless-telephone contact and they were prepared for it. Punctually at the time established on the chosen day the communication took place in perfect order. These were the first times that this happened in the whole world!

I must point out that in that era it was not yet possible to communicate via wireless-telephone from Hawaii to Italy. So it was an historical event and was much talked about. I remember Guglielmo's simple words when, looking at the world map on board, he showed me the distance separating those Pacific islands from Italy. "Look," he said, "Elettra is up there and we are down here." Saying this he opened and shut his hands to form a circle in a familiar gesture that he often made when he spoke about wireless communications throughout the world. We felt really far away from Elettra but what joy and deep emotion we felt when we heard her clear little voice. I recall Guglielmo's beaming satisfied smile. He was really happy. Once again he had benefitted from the results of his long and exhausting labours.

Immediately after hearing the goodbyes, I leaned my head on his shoulder and sobbed. These moments were always so upsetting. But I remember Guglielmo's sweetness and affection. He would tenderly dry my eyes with the large white silk hankerchief that he kept tucked in his blue jacket pocket and would say, "If you have to cry like this every time, I won't let you talk to Elettra again." But of course he only wanted to comfort me. I perfectly understood that with these wireless-telephone communications Guglielmo wanted to hear our daughter's voice and know how she was. It was a gift to me too to compensate me for the long separation from our child. Only Guglielmo was capable of procuring this joy for me. I wanted to be close to my husband always and not be a hindrance to him. Even so I could not hide how the long time away from our daughter made me suffer.

After our stay in Hawaii, the voyage aboard the "Chichibu-Maru" continued pleasantly. Before arriving in Japan, there was the tea ceremony and then a gala evening in Japanese style. Later, during the evening, taking advantage of the cloudless sky as we sat comfortably on deck together with other passengers, we watched the most beautiful and original display of fireworks that I had ever seen. High in the sky, the coloured lights clearly formed chrysanthemums and other flower shapes making a great multi-coloured spectacle.

On our arrival in Yokohama, the first European to come and welcome us was the Italian Ambassador in Tokyo, Giacinto Auriti, who presented me with a bouquet of red roses. I remember how happy Guglielmo and I were to meet an Italian dignitary after such a long time. Ambassador Auriti was very kind and pleasant to us during our stay in Japan and accompanied us on our visits to the various cities and places.

In Yokohama, before leaving the Chichibu-Maru, the local authorities organised a flight of carrier pigeons as a good luck sign for us. They put a pigeon in my hand and while I was stroking it, at the sound of a signal, the bird flew away followed by the rest of the flock. This scene took place in the presence of the press, including that of the Japanese and the cinema newsreel crews who, from then on, followed us for the whole of our journey in Japan. What a lot of newspaper articles, photographs and newsreels resulted! Guglielmo, stood smiling next to me. That attractive ceremony marked the start of our visit to the Land of the Rising Sun.

Supporting what I have just said, several weeks later in Shanghai I received a letter from my parents in which they told me they had seen us in a film of the pigeon ceremony shown in a Rome cinema. They had taken Elettra to the cinema to see cartoons and, suddenly, during the showing, this scene of the pigeons in Japan came on the screen. When Mummy and Daddy appeared how excited she was, shouting with joy! My parents too were happy to see our faces after several months.

As soon as we arrived by car in Tokyo from Yokohama, accompanied by Ambassador Auriti, Guglielmo and I were at once received in audience by the Mikado, the Emperor, at the Imperial Palace. This was an immense 17th century castle, with pagoda-style roof, surrounded by a massive moat crossed by a drawbridge. I remember that the Imperial Palace had been preserved intact since its construction. Inside the enormous rooms were sub-divided by marvelous screens and everywhere there were large Japanese vases decorated with precious paintings while the floors were covered with carpets and straw mattings.

Emperor Hirohito received us in a dignified manner in the throne room. Speaking in English, he expressed his great admiration for Guglielmo Marconi and conferred on him the highest honour of his country: the Grand Cordon of the Order of the Rising Sun, a decoration ornamented with beautiful enamels in which red and white predominated. Guglielmo was proud and happy to receive this recognition which few people had the honour to possess in 1933. The Emperor also presented him with a magnificent antique Samurai scimitar dating back to the 14th century with a very thin sharp blade.

In Tokyo, we stayed at the Imperial Hotel where all the foreign embassies including ours were located. We were guests of the great benefactor, Baron Okura, and of the Foreign Minister.

Our stay in Japan was organized meticulously with all details perfectly taken care of. As companion, I had a delightful and beautiful Japanese lady called Yori Takahashi who belonged to an ancient family and was married to a young diplomat, son of the Finance Minister. She spoke perfect English since she had been educated at the Catholic College of the Sacred Heart in Tokyo although she was of Buddhist religion. She was about my age, most intelligent and cultured.

Although many years have gone by, my memories of Japan have always stayed deeply impressed. Tokyo--almost completely destroyed by the earthquake of 1908--was already a modern city in 1933. The houses, built using anti-seismic techniques, were low and square in order to be safe in the event of a new catastrophe. Along the main street, the Ginza, there were large department stores in American style with big neon signs.

People told me Tokyo did not express the true character of Japan, but the ancient cities we had the pleasure to visit later did. In that nation, everybody wore the kimono: men, women and children.

In Tokyo, the personalities of the city and government had organised a big Japanese-style dinner in our honour. In the enormous dining room of the Maple Club, Guglielmo and I together with all the other guests were seated cross-legged on the floor, forming a large square. We too in keeping with local custom, removed our shoes. In front of us was a small lacquered table, flower-painted, on which were placed various bowls containing Japanese food of various colours, which we ate using chop sticks. We were also served sake, the national drink obtained from the fermentation of rise.

During the meal, some attractive young geishas about 15-16 years old, danced with delicacy, lightly touching the straw-covered floor. They were wearing splendid kimonos and embroidered obis. Their long black hair was arranged in complicated hairdos, held in place with long costly hairpins. These geishas moved and swayed harmoniously to the accompaninment of antique Japanese instruments. When the music stopped, they tripped lightly round us, repeating ritual gestures full of symbolic meaning. I still seem to hear the rustle of silk and see their graceful figures so full of femininity and fascination. I have kept a photograph of them as they served tea to Guglielmo and me at the end of that evening.

We also took part in many official dinners in Tokyo invited by members of the government, by personalities of the diplomatic corps and by old Japanese families--occasions which were always accompanied by speeches praising the genius of Guglielmo Marconi and with kind remarks addressed to me.

Before leaving, we went to the famous Kabuki theatre to see a performance of an ancient Japanese tragedy acted by the most talented actors of the day. They wore authentic ancient costumes and their faces were heavily made up so as to make the performance faithful to the era it represented in all its details as an art of an extremely high level.

Since women were not permitted to perform at the Kabuki, female roles were played by very slender male actors who dressed as women and spoke their lines in high-pitched voices. The Kabuki theatre was covered by a large semi-circular pagoda roof. Inside, the small boxes were packed with Japanese men and women all wearing kimonos, sitting cross-legged on straw carpets. This multitude of spectators watched in silence for many hours, engrossed in the enactment of the tragedy.

We too, the only Europeans present, had to sit on the ground. We did not understand the language, but it was explained to us that the meaning and the value of the text was comparable to ancient Greek tragedies. After the performance, Guglielmo and I asked to go backstage to congratulate those extraordinary actors, so talented and famous. Some good photographs were taken of us with them and I have kept them carefully in memory of that occasion.

After Tokyo, we travelled around for some time, mostly in trains which were comfortable and modern, visiting the main cities of Japan. We crossed countrysides filled with rice fields, tea plantations and extraordinary, deep-coloured flowers, so different from ours.

Kyoto, the ancient capital of the empire for more than a thousand years (from 794 until 1868) which had remained unchanged for centuries was our first stop. I really find it impossible to describe all the small streets, the tiny pagoda-roofed houses and the little shops crowded with kimono-clad people. Babies were carried on their mothers' backs, a habit which has now become common in Europe. In Kyoto, cars were very rare, rickshaws were used instead pulled by very speedy coolies whom I will mention again later on. We, among the very few Europeans present, seemed out of place in that so characteristic environment. In 1933, diplomats and missionaries were the only ones to undertake the long journey to the Far East.

But Guglielmo had noticed, not without a feeling of pride, that often in the visitors' books the signature that preceded ours was that of H.R.H. the Prince of Wales, later Edward VIII who was then a young officer in the Royal Navy. He had made the round-the- world trip just before us on board an English warship.

In the Kyoto hotel where we stayed naturally we had to take off our shoes before entering our rooms. The floor we walked on was covered with a very soft straw carpet. All the furniture seemed in miniature. The beds were so short that we were not able to stretch out our legs comfortably. The furniture, made in exotic wood, was of really Lilliputian size and the rooms too were small with thin walls. Room service was excellent. In the evening, after ten o'clock, we used to hear the water running in the rooms next to ours because the Japanese always had hot baths before going to bed.

Then we went on to visit the beautiful temples of Nikko, hidden in a maple tree wood in reddish autumnal tints, and the enormous Buddha of Kamakura, the Shintoist temple of Ise where each emperor went to render homage to the sun goddess Amaterasu who, according to Shintoism, was the founder of the imperial dynasty. We saw the castle of Nagoya, built in 1612, and a display of chrysanthemums of really indescribable shapes and colours.

We admired Fuji Yama, the sacred mountain of the Japanese. It was a splendid sunny day and I can still see it as I did that morning set against the backdrop of the blue sky with its permanently snowy summit. The local people who went with us on the excursion told us we were very lucky because it was most rare to see their volcano in such splendid form and this would have brought us good luck. Then, silently and in a dignified manner, they made many deep bows to their sacred mountain. The countryside around Fuji Yama was fascinating. In the little inlets, the sea lapped the coastline covered with low-growing vegetation and characteristic small pine trees.

By train, for hours on end we crossed rice fields and saw many Japanese of both sexes working there wearing straw coolie hats and with the water up to their knees. One of the most interesting and modern things we saw was the cultivation of pearls. Guglielmo and I had been invited by a Mr. Mikimoto, who was considered the inventor of the cultivation technique. Formerly a fisherman and already elderly at that time, he then became a real authority in his country. He took us to visit his factory of cultivated pearls at Agobay, south of Kyoto.

We admired the young Japanese girls, called "ama", who dived into the water to insert the foreign body into the oyster shell. In order to eliminate this, the oyster covered it with various layers of mother-of-pearl thus, unknowingly, creating a shiny pearl. These "ama" spent their youth in the cultivation of the pearl-making oysters.

We dined with Mr. Mikimoto in a pagoda built by the sea. He offered us fried brains and to my great amazement I found some beautiful pearls on my plate inside the food. Mikimoto, and Guglielmo who knew about this, were most amused at this. What a lovely surprise! It was an elegant way to give me a gift. In saying goodbye, Mikimoto also gave me a string of pearls to wear round my neck, each pearl being perfect in colour and shape.

A few years later, after Guglielmo's death, Mr. Mikimoto who still remembered my visit in Japan sent me a red Moroccan leather case containing a splendid string of black pearls. I so appreciated his gesture of participating in my mourning.

Nara was the next city on our journey; like Kyoto, it too is ancient built entirely in Japanese style. I still seem to see its park with its pagodas and the little lake covered with lotus flowers. On the little wooden bridges, kimono-clad ladies with their typical black bouffant hairdos held in place with long hairpins and carrying little parasols moved gracefully taking tiny steps, smiling and fanning themselves. How pretty they were! Today a scene such as this can only be admired on artistic porcelain or large ancient screens. But the real attraction of the park at Nara were the deer, still today considered to be a sacred symbol. I have kept photographs in my album on Japan that show us together in the midst of the greenery while we stroke the deer and offer them biscuits.

Then it was on to Osaka where we went expressly to visit the wireless centre, the most important one in Japan. Guglielmo was impressed by the perfection of the building and its equipment. In 1933, the Japanese had already set up the beam system. In the flat countryside two hammer-shaped towers stood out for the short-wave and micro-wave transmissions, the new system created by my husband and already in use in Great Britain and the United States. Everywhere we were surprised to notice how rapid was the technological progress of that ancient race. Also in Osaka Guglielmo had the immense satisfaction to see one of his more recent inventions applied.

Among other cities we also visited Kobe and I would still have much to tell about all these places and the persons I met together with my husband, but I prefer not to wander on at greater length.

I remember the look on Guglielmo's face when we left Japan. He was surprised to notice the strength of that great population which he connected to military organization and to the profound patriotism of men ready to fight spartanly until death in order that each one of them might inherit the ancient warrior spirit of their ancestors. When we reviewed some troops in Tokyo--this event is also featured in a "Luce" newsreel--Guglielmo was struck by how well equipped they were. This convinced him that the Japanese were capable of learning from the example of other countries and in this respect their army was modelled on that of the Germans and their navy on the British Royal Navy. They knew how to fight courageously to the point of sacrificing their lives, and their loved ones were likewise capable of this. In this connection, we were told about something that happened during the Russian-Japanese war of 1905 when a Japanese general had to go to war although truly anxious about leaving his wife alone. On saying goodbye, she realized his anxiety and a little while after his departure fearing that the general would not be able to fight with the courage and faith necessary to completely do his duty, she decided to commit suicide by hara-kiri. This was not an invention, but something that really happened.

My husband--who all his life has had a deep, sincere love for Italy, his native land--was struck with admiration at the action of that young Japanese woman.

With regard to love of country, I wish to remind people that Guglielmo Marconi, after his first exceptional invention of wireless and after having obtained the patent in England with the historic number 7777--known in English as the "four sevens"--desired that Italy too should take advantage of and make immediate use of its benefits. Therefore, without any financial compensation he transferred the patents to the Italian army and navy.

When we were about to leave the country by train from Shimonoseki station, we felt emotional and moved on saying farewell to the Japanese authorities and our friends. As we left, from the train window we could see the people in the crowd waving with the Japanese flag in one hand and the Italian flag in the other. Yori Takahashi, that dear young lady who had kept me company during the journey, was also there, a slim elegant figure in her kimono. She dried her tears on a little handkerchief convinced that she would never have seen me again. But, ever since, she has always sent me her greetings, as I have to her, with the help of Japanese ambassadors in Rome or through American diplomats.

46. Marconi with his wife Maria Cristina in Tokyo, Japan, November 1993, being served dinner by colorful *geisha*.

47. Marconi and wife Maria Cristina with Italian Ambassador, Giacinto Auriti together with Japan's greatest actors of the Kabuki theater--Tokyo, November 1933.

48. Hong Kong 1933: Marconi and Maria Cristina having disembarked from the liner *Conte Rosso*, accompanied by local dignitaries.

# KOREA AND CHINA

We left Shimonoseki together with our small entourage, as well as a great deal of luggage which included the many gifts and photographs which had been showered on us during the journey.

Guglielmo and I embarked on a small Japanese boat occasionally in service between the Yellow Sea and the Strait of Korea. On board was a Japanese captain, some sailors and a radio-telegraphist.

It was really a hellish night. At that time, there was no real passenger ship on that route. Guglielmo and I slept in the only existing minute cabin, with two bunks one above the other. There was a storm at sea and the wind howled. We had to hang on to the edge of the bunk in order not to be pitched out from one minute to the next.

At two in the morning, a man suddenly burst into our cabin in the dark--as if he were one of those pirates who infested that zone--allowing a recoiling wave to enter our cabin. We were terribly frightened but then burst out laughing when we realized it was the radio-telegraphist who had brought us a welcoming marconigram sent by the authorities of Dairen, the port in Manchuria in northern China, today known as Talien.

Our stay in Dairen was interesting and instructive. We stayed in the best hotel, the weather was fine and the sun shone. The buildings, in Russian style, built by the sea, gave the town a certain coquettish and, at the same time, dignified air. Even if since 1904 it had belonged to the Japanese, one could still detect the influence of the Tzarist domination in the architecture and general atmosphere. That territory had been the theatre of the terrible Russo-Japanese war which had lasted a year.

We were able to visit the fortresses, the battlements and the battlefields involved in the taking of Port Arthur, seized by the Japanese in 1905 after a long siege. We were accompanied by some retired Japanese generals who knew the zone perfectly since they had fought there as young officers.

Although many years had passed, those elderly military men in their halting English were able to give such an extremely vivid description of the events that we seemed to relive the famous, historic battle. Guglielmo remembered having followed the various phases, twenty eight years earlier, as published in newspapers of that period.

The visit to Port Arthur and its surroundings was most interesting. Here, too, the signs left by the occupation by the Tzar in 1898 were to be seen everywhere.

Guglielmo and I were most anxious to reach China, crossing Korea from the south. So we decided on this itinerary and we had the good fortune to see the authentic, ancient Korea, before it was devastated by the war. What a different world! From the port of Fusan, today Fusan in South Korea, for various days we travelled along the Korean railway line in a comfortable first-class wagon-lit. We were the only Westerners on that train which was very long and crowded and we had to go the length of it to get to the restaurant car. The Korean passengers slept in couchettes. What a lot of Oriental heads peeped out at us from behind the little curtains and we would smile back at them as we passed.

At dinnertime, we ate almost exclusively rice and hard-boiled eggs. Guglielmo was amused to note Bel Paese among the cheeses and Strega among the liquers. What a pleasure it was for us to find the coloured map of Italy, so far away at that moment, on the cheese label!

We crossed the flat Korean countryside in which the brown of the earth dominated, here and there were ancient villages. The low houses had grey clay walls which gave equally good insulation against both the heat of the summer sun and the winter cold and snow.

The train was rather slow thus we were able to observe distinctly, as if they were paintings, those villages and their inhabitants wearing the traditional Korean dress. The men wore stiff black felt hats and the women full skirts of coarse white canvas, gathered at the waist, seven metres being needed to make each one we were told. Next to the houses, there were clay fountains where the women washed these skirts and then spread them out to dry.

In the autumn sky, the sunsets of the Far East were coloured blue and orange and the chilly air announced the approach of winter. When we reached Seoul, the capital, we stayed two days lodging in the best hotel where, at last, after a long time, we were able to enjoy European cuisine. In the mornings, Guglielmo and I had a good English-style breakfast and during the day we enjoyed excellent French food.

We visited some temples and the most interesting museum. I was given a beautiful, antique Korean vase of a pale green colour. In the zoo, we saw a magnificent example of an oriental tiger.

In 1933, Southern Korea as far as Seoul was under the domination of Japan which had occupied it in 1910. The fear and preoccupation of the Korean population was, unfortunately, evident.

We were invited to the governor's palace and received by the Governor, a high-ranking Japanese general. He was most kind and hospitable towards us but aware of the strength of his country. One could feel that Japan intended having supremacy over the nearby China.

From Seoul we took the train heading northwards up through Korea and so entered China from Tientsin. In eighteen hours we could have reached Vladivostok where the Trans-Siberian railway ended. Guglielmo and I studied the itinerary on the map. At that stage, we were tempted to pay a visit to Russia, but preferred to dedicate the most time possible to seeing the great country of China.

Tientsin was the first stop in the Asiatic continent. Both the Italian and the local authorities welcomed us festively and filled us with attentions. Giglielmo and I were pleased to visit both the Italian and the foreign concessions, the Italian navy and the military forces. Also there was the battalion of the San Marco legion consisting of foot sailors whose task it was to protect our legation according to an agreement made with the Chinese government after the Boxer War of 1900. It was touching to see our flag in that faroff land and to hear the Royal March played.

We met various persons from the foreign concessions, French, English, in a pleasant atmosphere of great cordiality and friendship. How far away were wars in 1933!

Our journey continued across northern China. How exciting and joyful it was for Guglielmo and I to reach Peking! In December 1933, in that city there were only those Chinese who had survived the fall of the Empire in 1912. The few Europeans still there after the Boxer War told how there was not much left of what was there before. Anyway the atmosphere was still that of olden times. The Gates of Peking were still those of the past as was the Imperial Palace, the Temple of Heaven, the Imperial Winter Palace and the Summer Imperial Palace, the "Coal Hill" and the Jade Palace.

It is difficult to describe what the real atmosphere was like in Peking. It is said that the blue of the sky lasts for six consecutive months. The warm shades of the houses and buildings reminded one of the colours and light of Rome seen from the Pincio at sunset.

From the Astronomical Observatory situated up on the hill dominating the panorama, one enjoyed the view, the beauty and the mystery of Peking. The city was immense and at some points still surrounded by its ancient walls. From the main gates, the most important roads began and crossed the entire city. The rather low houses, built close together, had curved roofs covered with shiny tiles in which blue was the predominant colour. One after another, they formed an undulating line shining in the sunlight.

A much-travelled foreigner whom we met there pointed out to me that in the world there are three great ancient cities--Rome, Peking and Athens--that are alike morally-speaking because in olden times they have been the cradle of the major civilisation in the world. Probably for this reason, it seems as if they have in common the same atmosphere and the same colour at sunset: Rome from the Pincio, Peking from the astronomical observatory, Athens from the Acropolis.

Among all the Chinese we met, the inhabitants of Peking seemed the best looking: tall, slender, elegant and with neat facial features as if carved in ivory. Guglielmo was entranced by the children, finding them delightful with their pretty ways, their almond eyes so intelligent and quick. There were so many of them, hundreds of thousands, a really staggering number. The streets were crowded with these little children and a great crowd of men and women, all of them quite active in the cold climate of northern China.

They ran around on foot or sat in the rickshaws pulled by the barefoot and tireless coolies. Although it was winter, beads of perspiration ran down their bare legs. It really bothered us to see them but not so the Chinese who were used to making use of their services. With their cropped-off trousers and characteristic straw hat, they ran along like so many thoroughbred horses. When elderly, often they became ill with tubercolosis and sometimes they could be seen, collapsed in the street, next to their rickshaws. With great grief, I learnt that they were allowed to die like that.

Men wore a long straight tunic high at the neck and fastened with a long row of buttons. All of them were dressed alike, in indigo-blue to match the sky overhead. It was really something to see that numerous Asiatic population, all in blue, with their elegant movements, and their small, close-cropped heads, black and shiny, their facial features finely chiselled, always ready to smile.

The missionaries and the diplomats who were the only Europeans in all Peking, told us how difficult it was to understand the psychology of the Chinese, or to succeed in getting to know them intimately. Very sensitive but rather diffident, they did not show their feelings but hid them behind the so-called "Chinese mask."

Females wore rather wide trousers that narrowed in at the ankle and a jacket that closed high at the neck. As infants, the young generation had not had to endure the torment of bound feet, but the elderly women had difficulty tottering along on their tiny crushed feet. This cruel custom had been abolished twenty years earlier with the fall of the empire. From that time, the men no longer had pigtails and we only saw a few old people still attached to this old habit with their locks still in a long plait.

Guglielmo and I liked to wander about on foot among the people, stopping in front of the little shops selling porcelain and Chinese curiosities. We wanted to observe close at hand that population of a most ancient civilization which still, despite the passage of time, lived attached to traditions of a far-off age ignoring progress.

How different it was from Japan! That country had assimilated a lot from ancient China and the Europeans and had progressed rapidly.

Several times we went to Jade Street. It was narrow, with shops selling curios, porcelain vases of various epochs, shapes and sizes, but above all necklaces and art objects in jade and quartz, as well as ivory statues in all dimensions. But even so, it was a small amount compared with what you could find at the time of the fall of the empire. Before 1912, that street was famous throughout the world for the inestimable and incredible richness of an enormous quantity of objects and jewels of extraordinary beauty.

We were impressed by the magnificent astronomical observatory, designed and built in bronze at the end of the 1500s by the Jesuit fathers and even, so it is said, by the renowned traveller and Catholic missionary, Father Matteo Ricci. I seem to see once more the great twisting dragons, emblem of the Celestial empire. Guglielmo and I noticed in China so many scientific notions, so many profound traces left to posterity by the various missionaries in Peking, and above all by Father Matteo Ricci, the first of them to arrive in that city. In his first journey towards the Far East in 1579, he stopped at Goa in India. For thirty years he conducted the exemplary life of a missionary in the China of the Ming dynasty and thereby acquired a profound knowledge of the Orient. He spoke mandarin, a privilege at that time only of those extremely cultured persons, that divided the population in dominators and dominated. He was well known too and appreciated by the emperor to whom he offered his great moral qualities and his wisdom in an atmosphere of complete harmony. Matteo Ricci was a famous astronomer, geographer and mathematician. He arrived in Peking in 1601 and died there in 1610. His tomb is still venerated there by the Chinese.

Guglielmo and I visited the seat of the Apostolic Legation, the exterior of which was built in Chinese style. There, we met Monsignor Ildebrando Antoniutti, councillor to the Pontifical Legation, who later was made a cardinal. At that time he was a young prelate, enthusiastic about both his work and Peking. He knew the country's history, culture and art as, through direct experience, he also knew the Chinese people. The Apostolic Legation of Peking had a radio station unique in China, installed inside the nunciate in order to communicate with the Vatican. Even if I have already mentioned this, I like to remember that Vatican Radio was one of Guglielmo Marconi's important works, created and offered by him to Pope Pius XI in 1931.

Since we had been unable to communicate with Italy for several weeks, Monsignor Antoniutti allowed us to use that radio station to send a message to our faraway loved ones.

Unforgettable for me were the visits to the Imperial Palace together with my husband and accompanied by Monsignor Antoniutti. Guglielmo and I entered the famous Forbidden City through a beautiful gate, designed by the artist Pailou, in finely carved wood painted in gold and bright colours. We wandered around for some time and it was an enchanting visit. The high walls closed in the imperial palaces, with the rich marble staircases flanked with great dragons. Yellow and green predominated, these being the colours of the imperial family. We were able to improve our knowledge of art and history, ancient and complicated, of what the immense Chinese empire had been. Visiting the Forbidden City it was possible to imagine how marvellous Peking must have been at the time at the time of the great dynasties. In the huge atrium, there was the throne of the last empress, the terrible Tz'u Hsi, who governed China despotically from 1881 to 1908 and was still remembered by some. At a reception, we met a lady dressed in beautiful silk and wearing a precious jade necklace. She told us that when young she had been lady in waiting to the empress.

I still remember vividly the magnificence of the imperial palaces, the perfection of the Temple of Heaven with the cylindrical roof (unfortunately there was little left in the interior).

Almost all the works of art had been looted during the Boxer War of 1899-1900.

Guglielmo and I were fascinated by the quite unique atmosphere. The constantly blue sky and the green and yellow colours of the enamelled dragons which sparkled in the sun are still today clearly impressed in my memory.

We were almost alone in that now silent, great Forbidden City, seat of the imperial past. In that moment it was marvellous to think that already several centuries earlier human genius had created and built such grandiose works, constructed magnificently with perfect architecture.

My husband admired everything in silence and I was really happy to have undertaken that journey with him. We were really united in our feelings, we had the same tastes, we were both interested in art and history, because we both loved what was beautiful.

Accompanied by Monsignor Antoniutti we also visited the Catholic University in Peking, a large building in Chinese style. It was an unforgettable occasion. The Chinese students welcomed us with such affection and enthusiasm that we were deeply touched. There were several hundreds of them and can still see them smiling, tall, slim and elegant in their indigo-blue tunics as they gathered eagerly around Guglielmo. They wanted to touch him to express their admiration, shake his hand, ask for an autograph. For them it was incredible to be able to meet in person the inventor of wireless, the genius who had come from faroff Italy.

In the university garden, my husband made a short speech to the students in English, encouraging them in their studies and reinforcing their ideals. He was greatly applauded. Guglielmo Marconi had made them happy.

In Peking we were guests of the Italian Minister Boscarelli in the Italian Legation which was considered to be the most beautiful of all the foreign legations. At either side of the main entrance there were two great characteristic Chinese bronze lions which had been removed from the Imperial Palace by Italian sailors taking advantage of the great confusion that reigned during the Boxer War.

When we returned to our suite in the evenings we had the pleasant sensation to be back in our own country and finally we could enjoy the specialities of Italian cooking. The service was perfect carried out by a score of Chinese servants. I remember how well they ironed the lace and ribbons on my lingerie and how well starched were the shirts and dickeys belonging to Guglielmo who was very particular about such things. They made life easy for the diplomats and occasional guests being good workers, devoted and honest. They worked quietly, always smiling, and grateful for very small tips.

Guglielmo and I were invited to various receptions and dinners given by the heads of the missions and by representatives of the Diplomatic Corps. All were fascinated by Peking and were happy to reside in this great Far Eastern country.

The Chinese personalities were most intelligent and cultured, most having studied in Western universities, in England, the United States, or at the Sorbonne in Paris or Lovain in Belgium.

We noticed during our long journey around the world that the only language spoken by everybody was English, even the servants knew how to make themselves understood with their characteristic pigeon English.

In China, the ladies wore the traditional long slim-fitting dress, embroidered in silk, with the high neckline and side slit up to the knee. This is still the dress worn by the women in the island of Formosa. In Peking, the most elegant wore fabulous jewels in antique jade.

Before leaving, we were keen to visit the Great Wall of China, that colossal means of defence against the Mongols and Tartars. From Peking, Guglielmo and I travelled by train to Northern China, in Manchuria and Mongolia. We went to the cities of Mukden (today Shenyang) and Harbin (today Pinkiang) not far from the steppes still inhabited by the Tartars.

We spent two nights in Harbin, visiting some very beautiful temples even more ancient than those of Peking. It was December, very cold and snowing, with the temperature several degrees below zero. The British Ambassador in Peking, knowing the cold climate of northern China, had kindly lent Guglielmo his fur coat and his cap with ear flaps. I had my mink coat. The cars were heated with warming pans.

Harbin was a small town, solitary and sad while Mukden was very old and interesting. From the year 1644 it had been the cradle of the Celestial Empire in that it was the land of origin of the last imperial dynasty, that of the Ch'in. But at the time of our journey its splendour had diminished. I recall that we were told a short while before our actual arrival in Mukden in 1931, that the new emperor Pu-yi had been installed by the Japanese in a puppet government in Manchukuo. As a young boy, he had been the last emperor in Peking where he had reigned until 1912.

But the conservative Chinese did not have much faith in this event as was obvious from an observation of the political events that took place after our stay. In point of fact, Pu-yi remained as emperor of Manchukuo until 1945, the year in which he was arrested by the Soviets and then handed over to the Chinese of Mao Tse-tung. To us foreigners, at that time, Communism seemed far away from China, but the Chinese had a presentiment about what might happen and were very much afraid.

The Manchurian countryside was covered with snow and the landscape had a desolate look with only a few isolated houses belonging to peasant farmers.

We returned to Peking after a few days because it would have been too risky to stay any length of time in those deserted and very cold places completely isolated from the rest of the world. In case of need we would not have been able to find even a minimum of medical assistance and the Chinese pharmacies only stocked local products. Everything was completely Mongol.

Guglielmo and I were invited again to numerous dinners and receptions in the Chinese homes of local personalities. In their drawing rooms we admired beautiful painted panels, antique dinner services in fine porcelain, marvellous vases--some in gold, others in deep blue--of various epochs and many masterpieces in ivory and jade. The meals were exquisite. The best chefs in Peking prepared characteristic dishes for us such as tiny eggs that had been preserved in some mysterious way thirty years earlier, as well as swallows' nests, shark fins, bamboo hearts and other specialities. Once we were shocked to see even a dog served on a great dish. At the end of the meal one drank a hot consomme' made from orange peel while with meals Sake and local liqueurs such as "Rosebuz" were served.

Guglilelmo and I left Peking with great regret knowing that the possibility of our ever returning there was very remote. At the station, with our little entourage, we climbed onto the famous train then called the Shanghai Express. It was a long train pulled at speed by a continually puffing engine. We travelled for two consecutive days and nights in a wagon-lit, crossing endless stretches of flat fields that were little cultivated and through which wide muddy-yellow rivers ran. In those cultivated fields, soya was the most frequent crop--much used in China for the making of oil--and there were vast expanses of paddy fields.

For hours at a time we would see nobody, only little heaps of earth where the peasants were buried in the countryside when they died. The mortality rate was high above all in summer because of the frequent cholera epidemics. At that time China had a population of about six hundred and fifty millions but nobody had ever managed to make an exact count. It was said that an incredible number of Chinese were born and died within the first year which made any form of registration impossible.

As I have emphasized, preciseness was one of Guglielmo's main characteristics and he was therefore interested in statistics. He was amazed and unable to understand how the Chinese had not been able to organize themselves in order to know the exact figure of their population even if it were so numerous.

During our two-day train journey we saw in the distance groups of low-built houses that seemed to form a big farm. In reality these were nuclei of old families of peasants whose life and psychology are so well described in Pearl Buck's book "The Good Earth." In the main house which was in the centre lived the patriarch, the head of the clan and, around him, lived his children, grandchildren and great grandchildren, all with their wives and offspring, altogether forming a village of up to two hundred persons. They were born, lived and died within the same family nucleus, tenaciously attached to the old traditions being truly patriarchal familes. Medicinal herbs were used as health remedies, rice was the main food and they lived fortified by their absolute religious faith--they were almost all buddhists--reading and respecting the sayings of Confucius.

Usually Chinese peasants to go from one village to another travelled long distances following the direction of the rivers. There were very few railway lines, the most important being that used by the Shanghai Express. Guglielmo and I would have really liked to know more about the usages and customs of these Asians, so different from ourselves. Therefore, we were interested in getting to know and talk to the various local personalities.

Arrived in Shanghai, we stayed at the Cathay Hotel as guests of the owner, Sir Victor Sassoon, whom we knew well from London. Pleased to have us, he put an excellent suite at our disposal which reminded us of the one we had stayed in at the Savoy in London. It was really comfortable and luxurious with perfect service and good food.

In 1933, Shanghai was the embodiment of modernized China and was much influenced by the United States. It was the only city in which there were a few, not very tall, skyscrapers as well as busy traffic and well-lit shops with neon signs in Chinese characters. There was intense night life in the cabarets and night clubs where the dances fashionable in Europe were already known. The Chinese women were modern, attractive and elegant in their close-fitting dresses in beautiful coloured silks.

What a difference from Manchuria and Peking! It seemed like another world! Guglielmo and I considered ourselves to be privileged to have known the old true China before having arrived at Shanghai.

We visited the university where we met the chief dignitaries and were received with a warm welcome and much admiration. There too, the value of Guglielmo as a scientist was recognized.

Still in Shanghai, we went on board the destroyer "Quarto" anchored in the port. I remember our excitement as well as that of the captain and crew at our meeting. The "Quarto" was the only ship of the Italian navy in service in the Far East.

From there, Guglielmo and I went on an official visit to Nanking, the capital of China at that time and the seat of the government, while the consulates of the various nations were located temporarily in Shanghai. The Italian consul at that time was Filippo Anfuso, who was still very young. Our car together with that of the dignitaries accompanying us formed a long procession which attracted the attention of the passersby.

Guglielmo and I were received in the palace lived in by the head of state, the president of the Chinese Republic of those days, Chiang Kai-Shek, whose name means "hard like mother of pearl." Our welcome was grandiose and solemn. Chiang-Kai-Shek, wearing a black silk mandarin tunic, was surrounded by members of the government--among whom was the Minister of Finance H.H. Kung, a direct descendent of Confucous and cousin to Chian Kai-Shek. The president talked at length with Guglielmo through an interpreter since he knew only Chinese.

Present at the evening reception that followed was the enchanting wife of Chiang Kai-shek, Madame Sung Mei-ling, whose name meant "beautiful soul." She was a beautiful young lady, elegant and intelligent. She stood out for the cleverness and fascination that made her personality so attractive and celebrated!!! During that meeting, I certainly did not imagine that I would have had the great pleasure of meeting her again not only in London but also on the island of Formosa. Many years later, in 1956, I went to nationalist China as a guest of president Chiang Kai-shek, accompanied by my daughter Elettra who had always been so fascinated by my stories of the Far East.

We went to Formosa by air on board a DC6 of Pan American. Leaving from Ciampino airport in Rome the flight took about 48 hours with various stopovers, in Beirut, Karachi, Calcutta and Hong Kong. General and Madame Chiang Kai-shek welcomed us with great cordiality in their residence not far from Tai-pei, capital of Formosa.

Elettra and I admired the beautiful antique furniture on which were placed artistic ornaments in jade, ivory and precious Chinese stones. Our attention was attracted above all by some 18th century pieces in bamboo and by others in tropical woods such as jacaranda.

Dinner was delicious cooked in French style. The artistic table centre was composed of stupendous pink orchids arranged in magnificent silver holders while the precious and rare table cloth was of beautiful Burano lace worked in needle point had been specially chosen by Madame Sung Mei-ling as a compliment for us Italian guests. We were struck by the fact that we were served at table by the president's most trusted gnerals and colonels in order to guarantee major security.

Chiang Kai-shek and his wife told us how much they appreciated and admired our courage and determination in undertaking that long journey to accept their invitation to the Far East. We shall never forget the interesting conversations and the affection with which Madame Chiang Kai-shek spoke to us in her perfect English. My daughter remembers clearly the encouraging remarks she made to her regarding her future.

We were accompanied by the Ambassador of Nationalist China in Rome, Yu-Tsune-chi, who had been a good friend of ours for sixteen years. Among the guests were the Minister of Foreign Affairs in Nationalist China, George Yeh, who was ambassador in Washington when the islands of Quemoy were bombed by the planes of Mao Tse-tung. Also present at the dinner were various other members of the government in Formosa together with their wives all of whom knew English well. They were very intelligent, refined and cultured. Some of them had more than one university degree but even so modestly did not boast about these achievements. These Chinese who represented the elite of Nationalist China were characterised by their cultural and political qualities.

Going to the Island of Taiwan with Elettra, after having been 23 years before in China with Guglielimo, my husband, I was happy to see again some of the people I had met in Peking and Nanching. They had been younger then, of course, and later in Formosa they had reached high office after escaping the persecutions and after great suffering including the separation from their loved ones of whom they no longer had any news.

But now after this parenthesis, I shall return to the account of my journey round the world with Guglielmo. In Shanghai, we embarked on the Italian ship "Conte Rosso" belonging to the Lloyds' line. Once again Guglielmo and I regretted having to leave the Far East but we were homesick for Italy and above all for Elettra.

# INDIA AND OUR RETURN TO ITALY

The sea voyage to Italy on this modern and well-equipped ship took twenty-five days. The passengers came from different nations: Malaysia, Japan and India. Everybody was satisfied with the service on board the "Conte Rosso" and particularly pleased with the excellent Italian chef.

Life on board was enjoyable. Guglielmo and I made friends with those English passengers who liked to visit their dominions, most of them being former colonial governors and officials with their families and there were also diplomats.

The stop-over in Hong Kong was fascinating. We were invited to lunch by the British governor where we met the various local dignitaries including the well-known Jesuit Father Gezzi of whom it was said he had the gift of being able to forecast typhoons. For this reason he was kept in Hong Kong because it was thought that great disasters could be avoided by taking due precautions in time.

We visited the city with its typical districts of Victoria and Kaloon where we bought silks and souvenirs. We went to see the picturesque Aberdeen, the centre inhabited by fisherfolk who always lived on board their junks called "sampams." Hong Kong was so beautiful and elegant and in that period one could feel the opulence and supremacy of Great Britain. In the main square which gave on to the sea stood the bronze statue of Queen Victoria, standing up and holding the orb and sceptre, she seemed to welcome those who arrived by sea showing the power of the great British Empire. The city and the traffic were controlled by the police who were tall and elegant Sikhs wearing pastel-coloured turbans.

From Hong Kong, we resumed our voyage on board the "Conte Rosso" heading for Singapore. Before arriving there, we anchored in the Strait of Malacca, where we disembarked in the attractive port and did some shopping for little wooden objects made by local craftsmen.

We stayed for two days in Singapore where we spent some time in the magnificent Raffles Hotel built by Sir Stamford Raffles in 1819 when he liberated the island. Guglielmo attributed great importance to Singapore telling me that due to its geographic position it would have developed enormously in the future becoming a vital point for telecommunications throughout the Far East. He had forseen that using his invention it would have been possible to make transmissions with the other countries of Asia. Guglielmo held a conference in English at the University of Singapore and many persons present have long remembered the interesting speech made by my husband.

We visited the city and found delightful souvenirs in the markets crammed with exotic attractions. We were accompanied by the Italian consul who acted as guide also on various excursions by car to the surroundings. The wild jungle was very close at hand and we were attracted by the Malaysian fascination of that far-off oriental pensinsula. The British influence was very evident. The countryside grew luxuriantly around the beautiful country houses of the English colonisers, diplomats, army officers and other local dignitaries. The local people showed great kindness towards us.

In 1964, I was overjoyed to see Singapore again when with my daughter Elettra I went to Australia on board the "Guglielmo Marconi", the 45,000-ton turbine steamship of the del Lloyd Triestino making its inaugural voyage after it had been launched by me in Trieste.

I was the godmother of the ship, that bore my husband's name and when we arrived in Sydney Bay all the sailing boats came towards us to give us a warm welcome and many boats were following us. How hospitable and charming have been the Australian friends who have been always in touch with us since then. They come to Rome to visit me. Elettra goes to Australia often and loves it. There is the Club Marconi in Bossley Park, Sydney, which is developed in one of the biggest clubs in New South Wales which has also the Marconi Soccer team that almost always wins.

Resuming the story of my voyage with Guglielmo, after several long days of sailing we arrived in Ceylon, today known as Sri Lanka. We were enchanted by that marvellous island called "the pearl of the Indian Ocean." We disembarked in Colombo, the capital. The Italian consul-general was Menotti Garibaldi, nephew of our great Giuseppe Garibaldi. We were guests in his residence and he put his car at our disposal in order that we might have the opportunity to make various trips across the island. Along the beautiful coastline we saw the elegant houses built in English style, with well-tended gardens and verandas overlooking the sea.

Like me, my husband was also fascinated by the extraordinary sunsets, vividly coloured, that we saw through the palm trees and the vast tea plantations. Tea production was their biggest resource and tea was cultivated everywhere, on the plains, among the palms as well as on the hillsides. Guglielmo and I would pause to watch the beautiful girls and women of various ages who gathered the tea leaves and who would smile at us sweetly.

In Kandy we visited the temple where the faithful say that Buddha's tooth is to be found. We went into the forests and admired the ancient Buddhist temples inside which there were still traces of coloured frescoes depicting the life of the prophet. The botanical garden was unforgettable, considered to be one of the most beautiful and important in the world, with its great variety of flowers, plants, palms and other exotic trees coming from various parts of the world.

Nuwara Eliya had a cooler climate located in the mountains at an altitude of two thousand metres from where we could see the waterfalls that freshened the intense green of the forest. We heard the sound of the running water, the source of wellbeing of that splendid island. We went up towards Adam's peak and from there were able to enjoy the spectacular sight of the sun going down into the sea.

Guglielmo much admired the colonization carried out by the English as in the dominions of Great Britain. Wherever they arrived, while leaving intact the local beauties and characteristics, they knew how to offer to all the visitors the necessary comforts of life, constructing those few grand Victorian hotels in enchanting positions equipped with all comforts. Ceylon had stayed intact in all its splendour.

We left that paradise with regret. Guglielmo would have liked to stay there longer with me in that enchanting atmosphere in the island of Ceylon but his affairs in Europe and the necessity to be back in London because of his duties as president of the Marconi Company were on his mind and he could prolong his absence no longer.

The "Conte Rosso" continued its voyage as far as Bombay, then a city truly majestic and regal in its greatness and dignity. Guglielmo immediately thought of Disraeli who on completing the British colonies in the Far East in homage to Queen Victoria named her Empress of the Indies in 1876. We visited this most beautiful city by car and the Indian authorities even allowed us to enter the famous Tower of Silence, normally forbidden to foreigners. This was a great, open, circular tower without a roof where by means of a chute all around the dead of the Parsee caste were deposited and then completely devoured by the vultures which flew in from on high. We were so aghast by the description and even more so, because we could hear the noise of the wings of these horrible great birds. But I will say no more about this local custom.

To distract us from this macabre atmosphere, we went to lunch in the elegant Polo Club in Bombay, immersed in greenery not far from the city. Guglielmo and I watched the game with enthusiasm. Even if we had seen other games with Indian players in London at the Ranelagh and Hurlingham clubs, none had been as lively as this one played by those dexterious, elegant, young Indians with their coloured silk turbans and riding quick-footed ponies.

We resumed our voyage towards Italy. The academician Tucci, an expert on India, was also on board.

We celebrated Christmas 1933 and New Year 1934 on board the "Conte Rosso" in full sail in the Indian Ocean. These were joyful, carefree days, under a tropical sun. There were dances and very enjoyable parties, all of course well-organized by the captain and crew on board ship, in such a way that everybody felt perfectly at ease.

We both were happy and satisfied during these last days at sea that separated us from home. I remember Guglielmo saying just before we arrived, "You see, Cristina, we have visited many countries and seen many populations close at hand but what has stayed most in my mind is that everywhere there is goodness; everywhere the greatness and dignity of man is recognized!"

And then, finally, we reached Italy after having gone around the world! We disembarked in Brindisi in January. Knowing that Guglielmo and I were on board, Admiral Prince Ernesto Dentice di Frasso came to greet us and we were his guests in his splendid residence at San Vito dei Normanni at the end of the Old Appian Way.

But by now all that interested us was to hurry to Rome. In great excitement, we whispered to each other, "How much will dear little Elettra have grown? While we gripped hands to hide our growing emotion, inside myself I was saying, "Elettra, Elettra, we're coming, we are coming back to you!" And finally we were in Rome where we re-embraced our loved ones. I have no words to describe the joy I felt in being once again with my parents and our daughter.

As soon as we reached Via Condotti, after an absence of five months, we saw Elettra who ran to us shouting, crying and laughing all at the same time. I can still hear in the depth of my heart, Elettra's screams of happiness and ours too in that long close embrace.

In the days that followed we told relatives and friends about our experiences on the trip. They all admired our courage because in 1933 to go by sea towards continents so far away meant being isolated for long periods with no possibility of communication, cut off from everything and everybody and sometimes in dangerous situations. Among our friends and acquaintances, Guglielmo and I were the first to have been around the world.

Thanking God, we noted with joy that during this marvellous and unrepeatable journey, among changes of climate and temperature, among risks of all kinds, we had always been in good health and together had faced every sort of situation, without incident, returning feeling younger than before.

I am sure that this long experience, so pleasant and interesting, was most useful in order to illuminate still more the brilliant mind of Guglielmo. He, with his penetrating intuition, knew how to use the forces of nature which revealed their hidden secrets only to him.

# BRAZIL 1935

In these memoirs of my life with Guglielmo, I must not neglect another great event which was the result of his work: the illumination of the statue of the Redeemer at the summit of Mt. Corcovado at Rio de Janeiro in Brazil.

In February 1932 we were staying with my parents in Via Condotti in Rome. My husband wanted to give a demonstration of the results he had achieved with his latest radio-telegraphic experiments on long distance microwaves, so he had the idea of making a transmission from my father's house in Rome to the far off city of Rio de Janeiro in South America.

I can still see Guglielmo's upright figure standing there with such dignity and assurance beside the table on which his apparatus was placed. I was beside him and on the other side stood the Brazilian Ambassador as a witness to the event. At the moment agreed upon with Rio de Janeiro Guglielmo pushed a button and simultaneously a myriad of light bulbs lit up, illuminating the gigantic statue of Christ which to this day dominates the great bay of Rio de Janeiro. Those lights so far away lit up instantly. It was magical! To give a better idea of the importance of this event I must point out that in those days even the fastest ship took at least two weeks to reach Brazil from Italy.

There was great excitement all over the world and the news was on the front pages of the Italian and foreign press. Guglielmo was really delighted with this new experiment which was completely successful. It had been his own idea to make the demonstration by lighting up the statue; he was encouraged by his great faith in God, Who he believed was the source of inspiration of his work and the true Author of his inventions. He was driven by his passion to overcome all distances, even those across the oceans. He thought that it would be impressive to see the statue of the Great Redeemer lit up at night on the summit of the Corcovado mountain overlooking the bay of Rio, ready with open arms to welcome the sailors who had survived the dangerous crossing of the ocean.

Three years later, in 1935, the President of Brazil, Getulio Vargas asked Guglielmo to pay an official visit to Rio de Janeiro. My husband accepted his kind invitation.

We spent the spring and summer on board the *Elettra*; it was a very busy period for Guglielmo who was engrossed in his experiments on microwave radio transmissions.

On 12th September, 1935, we embarked on the Lloyd Sabaudo ship *Augustus*. We left our daughter Elettra, who by now was beginning to complain about our absences, in my parents' care. The voyage was very pleasant. In the morning we often went for a swim in the pool and in the evening different entertainments were organized for us. After a few days we sighted the Straits of Gibraltar. My husband took me and my brother Count Antonio Bezzi Scali who came with us on the journey onto the bridge. Guglielmo took us up to the captain who had no difficulty in pointing out the details of the Spanish and African coasts, the bays and the little inlets; he could calculate the exact distance between one promontory and another, the different depths of the water as well as the strength of the currents and the wind speed. He also gave us a very clear explanation of the route we were to follow.

Gradually as we got closer to Africa the temperature rose by many degrees. When we arrived at Dakkar in Senegal we were woken at dawn by the shouts of the natives who had surrounded the ship with their boats. We disembarked early in the morning and were welcomed by the Italian Consul. He took us to visit the town which over the years had become an important colonial city. The port was very large and the roads were busy with traffic; we were struck by the impressive height of the Senegalese. We also visited a characteristic village of native fishermen.

The next day we resumed our voyage aboard the *Augustus*. It was terribly hot both during the day and at night; everyone suffered from the heat and complained about it. Guglielmo was the only one who showed no signs of discomfort. When we crossed the Equator the atmosphere on board changed completely. The celebrations went on non-stop. My husband took part in the general merry-making, dancing with me and joking with everyone. In the evening there was a gala dinner during which Guglielmo and I had the pleasant surprise of finding a cake decorated with a sugar replica of the *Elettra* on our table.

After two more very pleasant days sailing I stood beside Guglielmo on the bridge of the *Augustus*; the sea was calm and there was no moon. We seemed to be enveloped by the darkness. From far off we began to make out the lights of Rio de Janeiro, dominated by the gigantic illuminated statue of the Redeemer with arms outstretched which overlooked the city and seemed to welcome us. It was the evening of 24th September. At that sight Guglielmo did not say a word; he was silent, immobile, visibly moved. He stood there for a long time deep in thought and beside him I too admired the result of his experiment of 1932.

It was late evening when we sailed into the magnificent bay of Rio. In the meantime a violent storm had broken out but in spite of this we found a large crowd of people waiting for us. The authorities came on board the ship to welcome Guglielmo Marconi. We received the Governor of the city and the Italian Ambassador His Excellency Cantalupo in the enormous stateroom of the *Augustus* which was packed with about two hundred people including many photographers and journalists all impatient to interview my husband. The rain was still falling when Guglielmo and I stepped ashore. Oblivious to this, he talked to the people waiting there who applauded with enthusiasm and cheered him loudly.

The next morning Guglielmo and I were received by the Foreign Minister who presented Guglielmo with the highest Brazilian decoration which was seldom given to foreigners; the Grand Cordon of the Order of the Southern Cross. Straight after this a luncheon was given in our honour at the Italian Embassy in the presence of the most important local personalities.

We were staying at the Copacabana Hotel which was outside the city; from the windows of our suite we could admire the ocean with its huge waves that crashed violently on the beach. Every evening two excellent orchestras played in the ballroom of our hotel, alternating Brazilian rhythms and European music.

Our stay was really pleasant. We could enjoy the natural beauties of Rio and its surroundings; the city was rich in wonderful scenery. We drove about in an open car. I remember with great pleasure that people recognized Guglielmo when we drove past them and called out to him enthusiastically with warm greetings. There was always a crowd outside the hotel waiting to meet Marconi. We often drove along the most famous roads of Rio: the Avenida Atlantica was an avenue that ran along the seashore for about five kilometres, while the Avenida Rio Branco ran through the centre of the city. We were slightly worried when we noticed that all the cars went at top speed.

One evening we were taken to visit the most famous night clubs, the Atlantic Casino and others. The atmosphere was electrifying with the dance floors animated by an international crowd; we saw young people of every race and colour dancing with unrestrained rhythm to the sound of the wonderful Brazilian orchestras. Guglielmo thoroughly enjoyed our tour of Rio de Janeiro by night.

I also went with my husband to the minicipal theatre one afternoon to listen the great tenor Beniamino Gigli; there was warm applause from the crowd for Marconi and Gigli.

The 26th September was an important day characterized by various functions in our honour. In a solemn ceremony the University of Rio de Janeiro conferred an honorary degree on Guglielmo; then there was a big reception at the Senate Palace where Guglielmo was warmly applauded. After this we went to the Brazilian Academy of Arts and Science where to our pleasure we met the great academic Aloysio de Castro who was deeply attached to Guglielmo. He made a wonderful speech in Portuguese and Italian at the end of which Guglielmo was nominated an honorary academic. My husband replied with his characteristic composure, making his thanks also in the name of the Royal Academy of Italy of which he was President. All those present listened to him with great deference and admiration; I remember all their eyes were fixed upon him. For the occasion he wore the uniform of the Italian Academy.

We returned to the hotel where we had to make a quick change of clothes; we had been officially invited to inaugurate the

Tupy radio-broadcasting station, the most important in South America. The Director was Francisco Assis Chateaubriand who came from Bahia. His talent had made him one of the richest Brazilian patrons. He was famous because he owned an important chain of newspapers to which he himself contrbuted articles with great intelligence and professionality. In fact, on the day of our arrival he wrote an article about my husband entitled "O Musquetero de l'Azul". (Oh Adventurer of The Blue). Government, industrial and scientific authorities were present at the inauguration. Guglielmo made a speech which was broadcast by radio all over South America.

Among the various receptions given in our honour the most important was the gala dinner which took place that evening, organized by the Foreign Minister in his magnificent residence at the Itamarati Palace to celebrate the birth of the new Radio Tupy. Guglielmo and I admired the precious Portuguese-style antique furniture and the magnificent floors in the vast salon. Passing through the french windows made of fine tropical wood we found ourselves on a delightful patio. We walked amid large palms surrounding a little lake where white swans swam.

Toasts were proposed in honour of Marconi by Chateaubriand and the Ministers. Guglielmo replied, thanking the Brazilian government for its hospitality and for the flattering speeches that had been made about him. Among other things he said:

"...This ceremony which has inaugurated a new and very powerful radio station is especially pleasing to me because it has allowed me to greet with deep feeling the great industrious Brazilian nation and the friendly nations of Latin America.

"...This solemn occasion is dear to my heart as an Italian because I can come into direct contact with the great masses of my hard-working countrymen who continue to love their glorious far-off homeland.

"...By now all the different applications of the radio are well-known and widespread, from the communication of ships and aeroplanes to one another and to the earth, to communications between the continents across the oceans and the mountain ranges...

"...One thing is certain; I can declare to you that from the day of my discovery to the present time I have never ceased to study and investigate those phenomena, anxious as I always am to wring some other secret from the enormous mystery of the universe...

"...By means of electric waves humanity does not only have at its disposal a new and powerful means of scientific research but is conquering a new force and using a new branch of civilization and progress that knows no frontiers and can even thrust us into that infinite space where never before have the heartbeat or any other sign of the activity or the thought of man been able to penetrate. This new force, which is playing an ever more decisive part in the evolution of human civilisation, is certainly destined to the general good by promoting the mutual knowledge and union between the nations, allowing us more and more to satisfy an essentially human desire, which is to be able to communicate among ourselves with ease and rapidity, annihilating that powerful element of separation which is called distance."

At the end of the dinner I was given a beautiful and original present: a column, almost as tall as myself, made up of three hundred and sixty-five orchids, all different, one for every day of the year. A grand ball followed, attended by all the most important families and the most eminent personalities of Brazil. The gentlemen wore their decorations, the officers were in dress uniform and the ladies showed off their most elegant evening dresses and their most beautiful jewels. They had a very high standard of living, given their immense wealth, which was unfortunately not well distributed among the population. They often travelled to Paris, New York, Rome and London to go shopping and they were always up to date on the latest fashions. They certainly lived a very free and easy life!

It was an unforgettable day. There were long articles in the Press about Marconi's visit to Rio de Janeiro. For several days photographs of Guglielmo, myself and our little Elettra were published on the front pages of all the newspapers.

The next day we went by motorboat around the bay of Rio. Guglielmo, who had been all over the world, told me that nature in that spot had created a really evocative landscape, unique in the world. I agreed with him completely as we admired the wonderful backdrop of mountains covered in brilliant green which were reflected in the water. We passed islands covered with a luxuriant vegetation; they looked like wonderful emeralds, perfectly cut, emerging from the sea. The entrance to the bay was dominated by Mt. Corcovado and nearby the island of Pào de Açùcar rose in all its pride. After about two hours in the motorboat we went ashore on the island of Pacquetà. Although small, it enclosed all the enchantment of a paradisiacal landscape. The rarest exotic flowers grew there; beautiful butterflies and brightly coloured tropical birds flew through the air. Our hearts were filled with happiness. On our return we were invited to a luncheon in our honour at the smart Yacht Club.

The morning of 28th September my husband decided to go on a trip by car. We visited the famous botanical gardens which had been created originally by the Regent Joao VI of Braganza and later embellished artistically and with very rare new plants by the Emperor Pedro II. We strolled along the avenues of palms which, tall and majestic, were lit up by the sun in a play of light and shadow that filled us with wonder at every turn; the rays of the sun filtered through the green of the countless plants that came from different tropical countries and made the jets of water sparkle in the imperial-style fountains. Next we drove towards Mt. Gavia; the road cut through the thick and luxuriant vegetation. We went through the *mato* the fascinating virgin forest. Little by little as we climbed we could admire the view of the sea, the bay and the city. When we returned to Rio de Janeiro we went to the race course to watch the Marconi Cup. The racecourse is one of the most beautiful in the world. I had the honour of presenting the cup to the winner.

The Brazilian government put a special train at our disposal for us to continue our journey to San Paolo. The railway stations at which we stopped were crowded with people who had come from every locality of the interior in the hope of seeing and applauding Marconi, the man who had astonished and revolutionised the world with his genius. Even in the early morning we were woken up by clapping and cheering from outside. Guglielmo would go to the window and smile and wave at everyone.

When we arrived at San Paolo, the Prefect of the city, Fabio Prado and other authorities boarded the train in a carriage we used as a sitting room to welcome us. Hundreds of people were assembled on the platform waiting to give Marconi a warm welcome. When Guglielmo and I got off the train we were surrounded by police who formed a cordon around us to protect us, even though with great difficulty, from the overwhelming enthusiasm of that delirious crowd.

We stayed at the Esplanada Hotel. After the enjoyable luncheon given by the Prefect, Guglielmo and I paid an official visit to the Governor of the State of San Paolo, Armando de Salles de Oliveira. He was pleasant but very dignified and even though he was very young he was already highly thought of both as a politician and because of his wide culture.

We only spent a few days in San Paolo which was not a large city in those days. I remember, however, that there was already the first skyscraper there, built by Francesco Matarazzo, who was one of the great pioneers of the country. While we were on board the *Augustus* we had enjoyed the pleasure of the company of the famous Count Francesco Matarazzo himself. He liked telling the story of how he had set out from Italy for Brazil when he was a young man, as an ordinary poor emigrant, leaving behind him his village of Castellabate near Salerno.

Guglielmo liked this man who had been active and resourceful all his life. In a very few years, thanks to his great courage and intelligence, he had managed to build up an enormous industrial fortune, becoming practically the founder of the city of San Paolo. When we met him, although he was eighty, he still personally ran many businesses of different kinds, being at the head of the largest industrial complex in South America. One evening Count Francesco Matarazzo invited us to a magnificent ball in honour of Guglielmo, where we enjoyed ourselves very much in a most friendly atmosphere.

During our stay in San Paolo we were also invited to luncheons and formal dinners by the Crespi, Pignatari and Prado families, the most important of the city; their luxurious houses were all built along the Avenida Paulista, a beautiful avenue in the upper part of the city. On these occasions the gentlemen wore decorations and dress uniform and the ladies decked themselves with tiaras and jewels to emphasise the importance to our visit.

One morning I went with my husband to the University. Guglielmo was welcomed by the students with the most touching enthusiasm. He listened to the young people's questions with interest and answered them carefully, immediately establishing a spontaneous and affectionate rapport with them. As usual he had words of encouragement for them and urged them not to let obstacles stand in their way.

While we were in San Paolo, Guglielmo gave me a wonderful surprise. He organized, right down to the last detail, a connection with Rome by radio-telephone; yet another proof of his genius.We were at a big official luncheon that day, guests of the Prefect of San Paolo Fabio Prado and his charming wife Countess Renata Crespi. Looking at his watch Guglielmo asked if they would excuse us a moment and took me into another room where a radio-telephone receiver had been installed. After we had talked to my father and mother Guglielmo and I heard the trill of our little Elettra's voice as she said breathlessly, "I've come up the stairs four by four to talk to you". How Guglielmo laughed at that! He enjoyed repeating this expression of our daughter's; it was so natural and spontaneous. He was delighted to have a long talk with his beloved little Elettra who he was looking forward to seeing again soon. I was overcome with emotion at the sound of my daughter's voice as she spoke affectionate words to me and sent me lots and lots of kisses. Elettra still remembers this episode clearly.

Our stay in Brazil was coming to an end. A two-hour journey in a special train took us as far as Santos. On our arrival at the characteristic tropical port we found the *Augustus* waiting to take us back to Italy. When Guglielmo and I got off the train we were met by enthusiastic cheers and applause; a Brazilian military band played in our honour. We were surrounded by the various personalities and our friends. In fact all the most important families of San Paolo were there to see us off and everyone came on board to say goodbye to us while the band of the *Augustus* played patriotic tunes. I was given quantities of wonderful flowers that filled our cabin and our apartment; a magnificent basket of orchids, a gift to me from the Italians living in Brazil, took pride of place in the main stateroom.

The ship drew away from the quay while the crowd continued to clap and cheer my husband. Guglielmo and I were really touched by their enthusiasm and we smiled and waved goodbye.

Many years later I returned to Brazil, invited by Senator Assis Chateaubriand, the President of Radio Tupy who had become ever more important over the years. Elettra and I met him in Rome at a reception at the Brazilian Embassy at the Doria Pamphili Palace. When Chateaubriand saw my daughter, who was so young and pretty, he was deeply moved because she had such a look of her great father Guglielmo Marconi who Chateaubriand had always admired enormously. He thought it was important that Elettra too should see the illumination of the statue of Christ the Redeemer at Rio de Janeiro and he wanted her to be the sponsor, in memory of her father, of Brazil's first radio-television station to be inaugurated at Ferrouphillha at the end of 1952.

After our return from a long journey in the United States in October 1952 Elettra and I left from Ciampino Airport in Rome on board a Paneiro do Brazil Airways plane and reached Rio de Janeiro after an eighteen-hour journey.

We were guests of the Foreign Minister and Senator Assis Chateaubriand at the Copacabana Hotel. We were welcomed affectionately by various personalities and Brazilian friends. I was happy to meet an old friend of Guglielmo's and mine, the famous academic Professor Aloysio de Castro; poet, philosopher and physician. He often accompanied us on our escursions and explained the characteristics and history of that great country which was then so wealthy. We went on delightful and interesting trips to Petropolis which had been built in the previous century by Pedro I of Braganza and to Teresopolis.

Elettra and I flew to the old Portuguese city of Bahia. We admired the stupendous sixteenth century church of St. Francis with its great twisted columns covered in solid gold and the Church of Good Hope built on a sheer cliff looking out over the Atlantic Ocean. We were among the first European travellers to fly to Bahia. In fact the city had only recently been connected to the capital by air. In those days only a few rare ships called there.

It was still a remote wild spot. Many churches remain as evidence of the flourishing period of Portuguese colonization and cultural dominance in the sixteenth century.

We flew with Assis Chateaubriand and his entourage in a government plane from Rio de Janeiro to Porto Alegre in the Santa Caterina region in the south of the country. From there we drove to Ferrouphilha which was quite close, where Brazil's first radio-television station had been built.

Elettra, who had been asked to inaugurate the new station, courageously made a fine speech in Portuguese. She spoke with emotion about all the great works of her beloved father. I too remembered the ties that had united Guglielmo to that great and noble nation of the South American continent. Senator Chateaubriand praised the inventions of my husband Guglielmo Marconi. Many articles appeared in the world press about this important event. At the end of these ceremonies we said goodbye to our dear friend Assis Chateaubriand, to the authorities and to the proud affectionate Brazilian people who were so full of humanity and joie de vivre.

We also went to Uruguay where we spent two very pleasant days as guests on the huge *fazenda* of the Bastos family who were important landowners. Elettra, a keen horsewoman, has many happy memories of the long delightful rides with the gauchos on the great grasslands of the Pampas.

And so the years go by; events follow one another, sometimes happy and sometimes sad. But you dearest Guglielmo will always be the light of my life; a light that I will keep burning brightly and transmit to our beloved Elettra. You will be remembered by humanity as a beacon whose noble works have made it possible to unite the peoples of the world and make it a safer place.

I have come to the end of the tale about my unforgettable, adventurous journeys with my husband Guglielmo Marconi. It all lives on in my mind and above all in my heart. As I share my thoughts with Elettra, who listens closely and with deep emotion, I relive all my life with him with intense feeling and fervent memories.

49. Brazil, Rio de Janeiro, 1935: Marconi with Maria Cristina surrounded by Brazilian dignitaries on the occasion of the inauguration of Radio Station TUPY.

50. Brazil, Rio de Janeiro, 1955: Marchesa Maria Cristina, Poet Aloysio De Castro, and Francisco Assis Chateaubriand (to the right of Maria Cristina) with other Brazilian dignitaries.

# VILLA REPELLINI The "PARABOLA" (later called the SATELLITE) 1934-36 THE *ELETTRA* AT SANTA MARGHERITA LIGURE

Having been an eye-witness to Guglielmo's various experiments I can testify to his being the creator of one of the most important and exceptional inventions of modern times. I am referring in fact to Radar and to the present use of the Satellite, which Gugliemo named the Parabola, as a means of transmission.

In April 1934 we were on board the *Elettra* anchored offshore from the port of Santa Margherita Ligure. I stress offshore because Guglielmo with his independent spirit did not want to be too accessible to the rest of the world while he was carrying out his experiments and I too enjoyed our isolation. Our little daughter Elettra lived happily on the yacht with us.

Apart from continuing his research on short and very short waves to perfect blind navigation, later known as Radar, Guglielmo began to work on a new invention using special apparatus that he himself built and perfected. The main and most important part was installed in his laboratory on board and the other in the garden of Villa Repellini, high up above the sea in direct view of the yacht.

I must mention here that the work done at Villa Repellini in the early 1930s by Guglielmo and his technical assistant Mr. Isted had resulted in the first micro-wave telephone system being installed between the Vatican City and the Papal summer residence at Castel Gandolfo which was inaugurated by the Pope in February 1933. Guglielmo gave the first important demonstration of this two-way communication to the Chief of the Italian

Telecommunications Admiral, Pession, and to the press in April 1932 between Villa Repellini and the promontory of Sestri Levante.

We went ashore from the *Elettra* and drove to Villa Repellini. Here, Guglielmo had set up a larger metallic semicircle, a little above his own height from the ground, with long needles about fifteen centimetres in length fixed to it and with a wire running to his headset. My husband stood there, listening and thinking; there was something magical about it all. Sitting there beside him I silently admired this new apparatus which Guglielmo touched with a satisfied smile, knowing that he had created another of his great and important inventions which would help to unite the peoples of the world. To my curiosity and questions he answered: "This apparatus is a Parabola (satellite dish) which will be very useful for universal communications". In fact, mysteriously for me, though not for him, from the radio station on board the *Elettra* by means of the above-mentioned Parabola Guglielmo created the satellite system, a fundamentally important scientific invention which is very widely used nowadays. There are many photographs of my husband with regard to that invention, while a famous illustrated weekly magazine published an article with a fine picture of Guglielmo Marconi standing beside his historic satellite dish.

While this new research was going on I saw that Guglielmo's brilliant mind was continuously producing incredible new ideas. He applied the electromagnetic waves in experiments in every field, even that of surgery. One day, during that last year of Guglielmo's life while we were on board the *Elettra* we were talking about the laser and he said to me: "You can't think how many serious illnesses could be cured by means of my waves". With regard to one of their medical applications, the ultra short wave scalpel, I should like to recall a declaration made by my husband which was published by "Sapere" a fortnightly magazine on science, technology and applied arts, Year 1, Volume 1, January 1935: "By means of this marvellous little instrument the surgeon can make a clean straight deep incision with hardly any loss of blood because the high-frequency current coagulates the blood in the blood vessels simultaneously with the progression of the incision".

When we were on board the *Elettra* he used to come straight from his laboratory and sit beside me on deck. I can still see him, wearing yachtsman's clothes, looking at the sea and the sky in the distance with a serene smile as if he were their master. He was happy and felt rejuvenated. He did not want to lose time but to hurry forward into the future; right up to the end of his life he continued to work incessantly; he always had new ideas and new projects because the world which only he had penetrated still seemed mysterious and to be explored more and more, ever more. His tireless insistent mind foresaw new and limitless possibilities. As usual in those moments as he talked about the future of his inventions Guglielmo made a circle with his beautiful hands. He was full of enthusiasm as he explained his ideas and projects: all linked to the possibilities of communicating and uniting the whole world by means of the waves that he alone had identified and picked up for the good of humanity.

At this point I should like to record an important event which took place many years later. On 1st February 1982 I was invited officially to London by Mr. Olaf Lundburg, the General Director of an important company, INMARSAT which produced systems for communicating with ships in the open sea via satellite. I was to inaugurate this new space communications company in a formal ceremony. Also present were Mr. Paul Robinson, the Managing Director of the Marconi Company and Mr. Peter A. Turrall, the Publicity Manager, with other representatives of the same Company, together with the most important representatives of various radio-communications companies and the principal British newspapers.

From the Equator--on the high seas off Bahia in Brazil--the captain of the transatlantic liner Queen Elizabeth II, Captain Peter Jackson, sent me a beautiful and very significant message recalling the achievements of Guglielmo Marconi. I replied with another message thanking all concerned for the historic event; having been an eye-witness I recognized the importance of my husband's last invention which he called the Parabola, while we were on board the *Elettra* at Santa Margherita Ligure. It was a great event which required a really exceptional organization of men and equipment. For the first the two messages were heard simultaneously by all the ships sailing in the Indian and Pacific oceans. Elettra, who was present, was also able to speak to the captain of a large Japanese ship which was crossing the Sea of Japan. I remember the profound emotion, evident in spite of their self-control, of the two ship's captains at talking to us via satellite.

After this important ceremony and a magnificent dinner that evening, the representative of the COMSAT, the most important American satellite corporation, said that he would like to invite me to the United States with my daughter. I accepted the invitation and two years later in 1984 Elettra and I went to Washington. We were guests at the L'Enfant Plaza Hotel from which we enjoyed a beautiful view over the Potomac River. It was next door to the COMSAT in an elegant locality very close to the White House. As we looked at the COMSAT building we could see the many satellite dishes installed on the roof.

The General Director, Mr. Joseph Charyk, and the other Managers had asked me to take part in the opening ceremony of this important Corporation in my capacity of Guglielmo Marconi's widow. My presence underlined the importance of this great conquest of mankind which had already been foreseen by my husband: communciation via satellites. My picture appeared on television together with that of Mr. Joseph Charyk and that of the organizer of the COMSAT. Simultaneously, thanks to those very satellites, some important personalities could be seen communicating from different parts of the world; among these, the British Prime Minister, Mrs. Margaret Thatcher from London. Ronald Reagan, the President of the United States of America, also took a great interest in this important event.

Elettra was present, excited and interested; she told me that she had been very much impressed by the sensational new transmission. She was in the big conference room of the COMSAT together with other personalities; large television screens had been set up so that my daughter and the other people present could enjoy the extraordinary event. I spoke first, emphasizing the importance of this new means of communication. On that occasion I also spoke with pride about the intense scientific work carried out by Guglielmo between 1934 and 1936 on board the *Elettra*, concluding with the invention of the Parabola, later called the Satellite. Everyone listened with interest because they knew that I had been an eye witness. Subsequently, Mr. Joseph Charyk described the exciting development of Guglielmo's invention.

Another demonstration of the importance of the satellite was given by the concert at the Baths of Caracalla in Rome on 7$^{th}$ July 1990. The three great tenors, José Carreras, Placido Domingo and Luciano Pavarotti sang, accompanied by the orchestra directed by Zubin Metha. Their wonderful voices were heard all over the world. That same night, so full of stars and music, I received many telephone calls after the concert from people in faraway countries; even from Australia. They were all moved and praised the great works of my husband, Guglielmo Marconi, whose genius had once more filled the world with joy.

51. Marconi, Maria Cristina and Elettra holding firm onto her father's arm.

52. Marconi with his wife, Maria Cristina and with their daughter, Elettra--London, 1934.

53. Marconi, his wife Maria Cristina and their daughter Elettra, while waiting for a private audience with Pope Pius XI--Rome, 1934.

54. Free spirited Elettra aboard the *Elettra*.

55. Marconi with his wife Maria Cristina aboard the *Elettra* on which was installed his newly invented microwave systems.

56. Marconi with his wife Maria Cristina in the living room aboard the *Elettra*. Behind them is the piano--his favorite instrument they both played together. Around the room were also photographs of many of their famous acquaintances.

57. Marconi and his parabolic antenna--Levanto, Liguria, 1935.

58. Aborad the *Elettra*, 1933-34. In one of his passes, Marconi is piloting his yacht between two buoys, using his "blind navigation" later called *radar*, in the sea of Sestri Levante, Italy.

59. Marconi and Maria Cristina on the terrace of Villa Repellini, where he built his parabolic antenna--Santa Margherita Ligure, 1935.

60. Marconi enjoying a broadcasting session in his own radio station aboard the *Elettra*.

61. Viareggio, 1935: Marconi, Elettra and Maria Cristina.

62. Rome, 1935: Elettra adoring her father.

63. Marchesa Maria Cristina Marconi with her daughter Elettra, 1935. (Photo Eva Barrett.)

64. Young Elettra peering through the wheel of her father's yacht.

# EXTRACTION OF GOLD FROM SEAWATER SANTA MARGHERITA LIGURE ON BOARD THE *ELETTRA* IN 1936

Now I want to write about a new invention of my husband's that nobody has ever heard about.

In the years following the illumination of the World Exhibition of Sydney in Australia from Genova in March 1930, we spent long periods in Liguria in the Gulf of the Tigullio on board the *Elettra*. On account of its geographical position, the area between Santa Margherita Ligure and Sestri Levante was especially suitable for Guglielmo's experiments. He devoted the last year of his life to research on blind navigation and the extraction of gold from seawater.

Untiring as always in his creative work, Guglielmo used to go to a little cabin in the stern of the "Elettra" where he carried out very special and important experiments to reach this new goal. He told me that the sea in the area around the Gulf of Santa Margherita Ligure was particularly rich in the precious metal. He used to repeat to me: "You have no idea how much gold there is in seawater. If I could find the way to extract it I could get more from the sea than is produced by the mines of South Africa".

I should like to emphasise that my husband was driven by patriotism, not self-interest; he thought that such a discovery would be of fundamental importance to relieve the financial problems of his beloved Italy. Scientist as he was, he knew the Italian territory well and knew that unfortunately it was almost completely lacking in precious metals, unlike other nations that were richer just because they possessed this resource.

Little Elettra and I were the only ones to know about these experiments and to be allowed into this special cabin. There was an acrid smell of acids in there. Complicated instruments were scattered all over the place: oscillators and fine copper wires, which we went together to buy in Santa Margherita, and other mysterious instruments. Great disorder reigned in this little cabin where not even the dust was ever disturbed; Guglielmo warned us not to touch anything: "There is order in my disorder", he said. He looked around him happily with a triumphant and allusive air.

Elettra was afraid to disturb him and watched him in silence with her big, intelligent blue eyes. She was proud of him. Although she was only a child, she understood the importance of his experiments and research; she realized that her father was accomplishing new works. And so, once again, we could admire the splendid result of a new invention of Marconi's: the extraction of gold from seawater. We were amazed; I remember that I exclaimed: "But how do you do it?". He replied, very calm and sure: "With my waves, with electric waves". He was quite right to call them "my waves" since it was he who had worked wonders with electric waves.

Elettra and I watched Guglielmo while he carried out the experiment, using a piece of coal. He enclosed this mineral in what looked like a little sack of very fine wire which he then left for a few hours immersed in seawater. He slipped the little sack from the porthole, suspended from a wire attached to mysterious electrical instruments. My husband explained to me that by transmitting energy from this apparatus he would provoke a chemical reaction in the seawater, so that the infinitesimal particles of gold suspended in the water would coagulate, unite and then, little by little, materialize in specks of that remarkable and fascinating metal. After a few hours Guglielmo pulled up the little sack: Elettra and I were astonished to see some thread-like fragments of gold which showed up on the dark grey of the coal.

At the end of the experiment the father, austere but kind as always, with the air of someone revealing a great secret, showed his little girl the particles of gold that shone so brightly, while Elettra uttered little cries of wonder and excitement, clapping her little hands; he smiled at her with amusement. Those two were really in tune with each other, in fact. It moved me to see them together because they were so much alike both in joy and in sadness. Even today, living with Elettra, I have the impression that my husband is always here with me. She is incredibly like her father.

Guglielmo, with his acute psychology, noticed that Elettra as she grew up was becoming more and more interested in his work; she was curious to know about it and anxious to understand. The constant company of a father who was a genius and his lofty and visionary conversation had influenced her child's mind, making her precocious for her age: a wonderful world opened up before her, infinitely superior to that of toys and dolls. I must say that our little Elettra was more attracted by her Daddy's experiments than by the usual games of her contemporaries.

My husband sometimes confided to me with emotion: " I shall be able to teach this child so many new and great things as she grows older! Together with you, she will be the witness and trustee of the future discoveries that I already foresee". Therefore, it was not difficult for me to teach Elettra to respect the memory of her father, Guglielmo Marconi, and to keep it alive for ever. In fact, she loves him so much that she feels he is always present; all the episodes of her life with her father and the affectionate phrases he spoke to her live on in her memory, never to be forgotten.

Guglielmo lost no time. One day, in the summer of 1936, he called me and asked me to come with Elettra into his secret cabin in the stern. We were anchored in the little port of Santa Margherita Ligure. My husband locked the door carefully and showed us three little glass jars, each with a lid. With a satisfied, self-confident air and the coolness he always assumed in very important moments, he said to me: "Look, Cristina: each of these three jars contains a sample of gold in three different colours: red, yellow and green". It was true! Through the glass I could see the shining little particles and each jar contained gold of a different colour from the other two. I was speechless for a moment, then I embraced my dear Guglielmo with emotion. I felt the responsibility of this secret of ours. Elettra still remembers all these details clearly.

When we disembarked from the "Elettra" in the autumn I tried to persuade my husband to put all his apparatus regarding the gold in a military chest. I would have helped him myself, ensuring that it was sealed up with the utmost secrecy, and we would then have taken it to Rome with us in our car: a big Lancia Dilambda by Pinin Farina. Guglielmo Marconi's wonderful apparatus would have been safe and hidden from prying eyes in his locked study in my parents' house. He, however, taking a firm and decided decision, refused. Apart from the usual risks of industrial espionage he was also afraid of a car accident; he did not want to reveal the secrets of this new invention of his.

To my great disappointment, before disembarking from the "Elettra"--which was laid up for the winter in Genova--I had to help Guglielmo while he completely dismantled all his apparatus regarding the extraction of gold from seawater; the three glass jars containing the three different-coloured varieties of gold also disappeared. Guglielmo had decided to put off this research, which by now showed positive results, to the following year, to the next season on board the "Elettra". To console me, he smiled and whispered to me: "Don't worry, Cristina darling, I haven't destroyed anything; it's all in here by now", pointing to his forehead!

It was one of Guglielmo Marconi's last, extraordinary discoveries; unfortunately he was not in time to communicate it to the scientific world with a conference, as he would have wished, because, to my great sorrow, his death came unexpectedly from a heart attack.

65. Marconi and Maria Cristina leaning on the railing of their *Elettra*.

66. Young, proud, and protective Elettra embracing her famous parents, aboard the *Elettra*--Santa Margherita, Ligure, September 1936.

67. Marconi, Maria Cristina and Elettra in the yatch's dining room--1936.

68. Elettra listening to her father explain his last invention on extracting gold from seawater using electric waves.

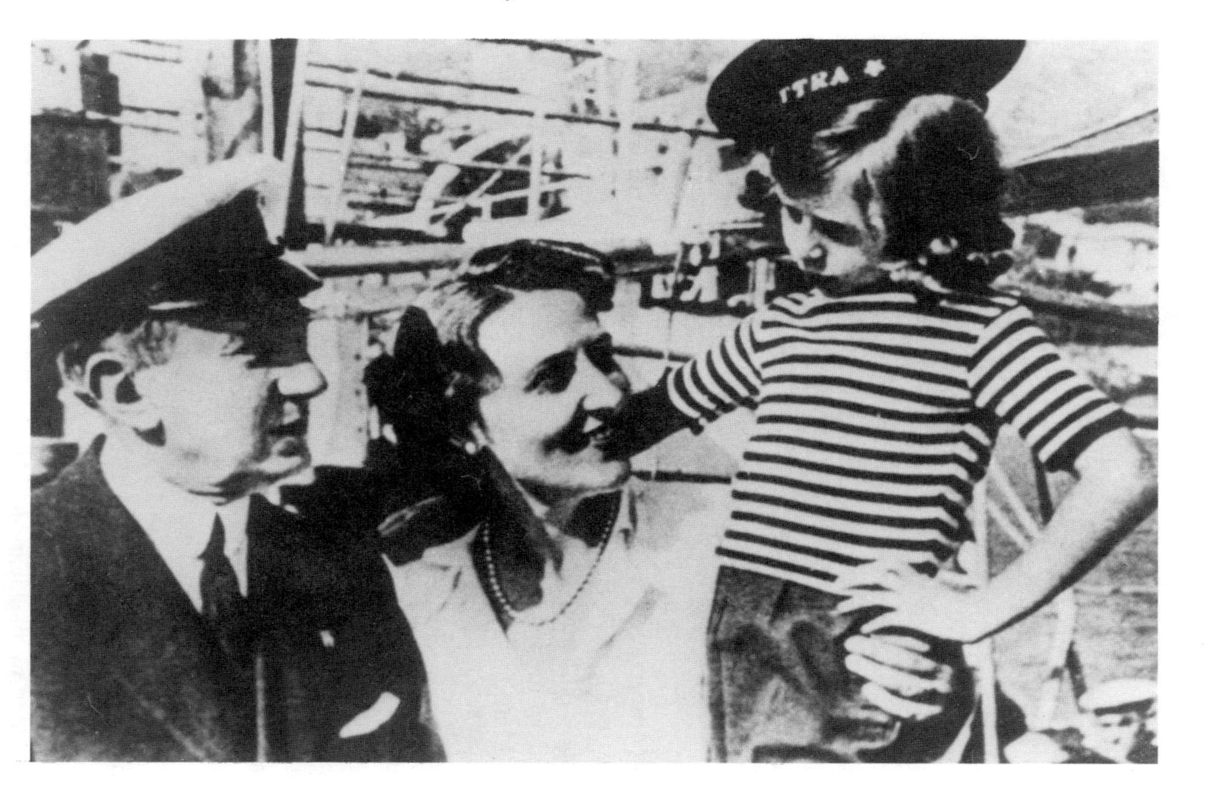


69. Proud parents admiring their daughter aboard their *Elettra*--Santa Margherita Ligure.

# THE DEATH OF GUGLIELMO MARCONI

My marriage to Guglielmo was a truly happy and Christian union. Guglielmo and I lived in perfect harmony because our love for each other was as profound as ever; I understood his determination to accomplish all the tasks, research and everything else he set himself to do with scrupulous accuracy.

His scientific experiments and his many activities and responsibilities in London as the President of the Marconi Company and in Rome as a Senator of the Realm, President of the Royal Academy of Italy and Founder President of the National Research Council meant that we were constantly on the move. My husband's many commitments meant that we had to undertake a series of journeys and voyages which were often tiring and uncomfortable; inevitable in those days since there was no rapid means of transport and airlines were not yet in service. Unfortunately, although he was strong and energetic, all this began to undermine his health.

I had always watched over his health but especially during the last period of his life I began to fear particularly for his heart. I took the greatest care of my dear Guglielmo and always looked after him with loving solicitude right up to the end of his life. I was motivated not only by the great love I felt for him but also because I felt it was my duty as the wife of such a genius; I had in fact married a man who was different from all others, one who had given and continued to give security and well-being to humanity.

Unfortunately there were some people who would have liked to come between us and spoil our happiness; they tried to do so and even wrote about it after his death in books which I do not wish even to name; but they never succeeded. Gazing into my eyes, Guglielmo Marconi often repeated, paraphrasing the famous Napoleonic declaration: "God gave her to me, let no-one dare touch her!" This phrase encouraged me and spurred me on to do my best for him.

During our two longest journeys we both suffered when we had to leave our darling little Elettra for several months. It was a great sacrifice for me but I did it for his sake and he truly appreciated it: we lived for each other. As I have said before, Guglielmo and I were very grateful to my parents, in whose affectionate care we left our daughter.

When we three were in Rome Guglielmo loved staying in the discreet family atmosphere of my parents'17th Century palace at 11, Via Condotti. They were delighted to have us to stay and to enjoy our company; we were often away for a long time abroad and especially in London, owing to Guglielmo's commitments with the Marconi Company. Many months went by before we could come back and see them again. I remember with great pleasure that there was always a mutual respect and a quiet harmony--so rare--between my parents and Guglielmo and staying with them gave us back the energy we needed to return to the very busy life we led during the rest of the year. It was actually in my parents' house that my beloved husband's heart stopped beating.

When we returned from our journey around the world Guglielmo's health was quite good on the whole. It was only in 1936, the year before he died, that we did not leave Italy because of my husband's health problems; having applied his mind and all his physical strength so intensely to his work, his heart was tired and he suffered from Pseudo-Angina Pectoris. He had the very best doctors wherever we were and in Rome his doctor was the famous Professor Cesare Frugone; with treatment and careful nursing, after a brief period of rest, Guglielmo's health improved and he returned to his normal life and his wonderful experiments.

I was careful that nothing should strain his precious heart; I watched over him anxiously to make sure that he did not get too tired and had all the care he needed; his strong constitution and great will-power helped him to overcome every difficulty. He could work for hours on end with intense concentration.

Nothing predicted his premature death. Not even Professor Frugoni was worried about his illustrious patient; he advised him and treated him skillfully but he did not foresee what was going to happen: in fact on the occasion of our last long voyage across the ocean to Brazil Frugoni reassured me, encouraging us to make the journey because he was convinced that it was safe for my husband to face it, although in fact it turned out to be very tiring for both of us.

There were no obvious outward signs of ill-health; in fact I can still remember the last days before he died. Guglielmo wanted me to go with him from Rome to Torre Chiaruccia near Santa Marinella to inspect the important experimental radio station there, before embarking on the *Elettra* for the summer season; he was feeling well.

The *Elettra* was anchored at Genova in the little "Duca degli Abruzzi"harbour with Captain Stagnaro in command; my husband had already given instructions to the yacht's radio-telegraphic officer for our imminent arrival. Guglielmo intended to start his experiments as soon as possible. The 20th July, 1937 was approaching: our darling Elettra's birthday. Our daughter was at the Astor Hotel in Viareggio with my mother who was taking care of her with loving dedication, helping me to bring her up well. Guglielmo and I were in Rome. The heat was intense. I was very worried about him as he still had a lot of work to do at the Royal Academy of Italy and the Marconi Company.

Our daughter Elettra's birthday was a very important date for my husband. The child was going to be seven years old and he wanted to spend her birthday with her as he had always done; this time it would be the afternoon before he could hold her in his arms. At all costs, however, Guglielmo wanted me to get to Viareggio the day before, that is 19th July, so that the child would not start the day of the 20th without at least one of us there.

It is painful for me to remember those moments and to write about this subject but I want to state with absolute honesty the true reason I was not at my husband's side at the moment of his death. On the morning of the 19th Guglielmo was well and appeared to be quite fit; we were only going to be apart for twenty-four hours.

In Rome, everything was already arranged for his departure the next day. He was not going to be alone in the house; my father, who knew about the state of his health, was there to keep an affectionate eye on him. In fact, at 5 o'clock, at the first sign of Guglielmo's illness, my father called Professor Frugoni who rushed there immediately and stayed with him until the end, doing everything possible to save him.

I shall never forget the dear, handsome figure of my husband on the morning of 19th July at the Termini Station in Rome. Guglielmo wanted to get into the compartment that had been reserved for me to make quite sure that nothing was missing for my comfort; he was not used to see me leave by myself, nor I to be without him. He held me in his arms and kissed me more tenderly than usual and told me over and over again to remember to give all his love to Elettra who he was going to see the next day. Then he got off the train and stood outside my window waiting for the train to depart. I can still see him, erect and dignified, dressed in light gray flannel, his Homburg hat on his head with the brim tipped down in front; he was leaning on his ivory-handled stick, a healthy colour in his cheeks. He looked at me with a gentler, more intense and more penetrating expression than usual: "See you tomorrow ! See you tomorrow!" he said; instead I never saw him again, alive.

I will always remember those last moments and his voice that whispered to me: "See you tomorrow...!"

I resign myself to the Will of God because I know how painful the moment of our final separation would have been for Guglielmo and me.

When the news of his illness reached me I felt as if I had been shot through the heart. Grief-stricken, I caught the first available train back to Rome and arrived in the Capital at about 7 o'clock the next morning. I hoped to find him still alive and to be with him when he drew his last breath. Unfortunately, my race was in vain: Guglielmo died at dawn on 20th July at 3.45 a.m. just when I was rushing back hoping to be in time to hold him in my arms again.

I shall never forget my father's pale, sad face when he met me at the Termini Station. I knew at once: my beloved Guglielmo was dead!

What a painful blow it was for me a few minutes later to see his still body lying in our bedroom at 11, Via Condotti; in his hands he held the rosary and the crucifix I had given him the day we were married. I kissed his cold forehead.

Sister Agnese, a Spanish nurse who was very attached to us, was a great comfort to me, together with my father, in such an immensely tragic moment for me. They described to me with what faith Guglielmo in the last moments of his life had repeatedly kissed the little crucific. He died with great courage and Christian resignation.

My father told me at once that in his last moments, when he was already breathing with great difficulty, he had called him to his side to ask him to be sure to send Elettra his telegram with his best wishes for her birthday and all his love.

Dearest Guglielmo, how dearly you loved Elettra. What high hopes you had for her. Sadly you did not have the time to enjoy seeing her grow up healthy and strong in body and mind just as you would have wished.

As I have already said, it was Elettra's seventh birthday that day and she was staying at the Astor Hotel in Viareggio; the next day she was to have gone to Genova with her father and me where we were to embark on the *Elettra.*

The moment she opened her eyes she had the unpleasant surprise of not seeing me in the bed next to hers like the previous evening. She found my mother there instead. The child instinctively understood that something terrible had happened. Was it perhaps the blood-tie that spoke? Some time later Elettra confessed to me that when she first woke up she cried quietly to herself and then she sorrowfully embraced her grandmother; she had realized immediately that her beloved father was dead.

Guglielmo lay in state at the Farnesina Palace. There were no flowers and no candles; it was austere and simple. People of every nationality and from every walk of life wanted to pay him their last respects. Those who were present said they had never seen anything like it; rivers of humanity came to pay tribute to the genius of the radio. Thousands of people followed the coffin and tens of thousands lined the streets of Rome for his funeral on 21st July, 1937. The solemn cortège crossed Rome from the Farnesina to the Church of Santa Maria degli Angeli. At 6 o'clock radio stations all over the world observed two minutes of silence. Marconi had broken the silence of the ether and for two minutes the world returned to that silence in his honour.

My husband, Guglielmo Marconi, should not have died so soon and so young. He had just had his sixty-third birthday. He had a very strong and healthy constitution but his heart was tired!

If he had lived longer, who knows how many other wonders he would have given to the world! Especially in the last months of his life he used to say to me: "You can't imagine, Cristina, how many hidden forces there are in the atmosphere which we don't know about and which would be so necessary to humanity". In his last conferences, he never failed to encourage young people to devote themselves to new research and never to give up but try and try again.

I always did my best to make Guglielmo happy, giving him my love and my constant care and attention. Our union and our great love were always perfect right up to the end of Guglielmo's life and continue to be so even now after his death. I never wanted to remarry and I have dedicated my life to his memory, comforted by the love of my daughter Elettra and my grandson Guglielmo.

70. One of the last photos of Guglielmo Marconi with Maria Cristina--Viareggio, 1936.

71. Senator Guglielmo Marconi--1932.

72. Signed photo of Maria Cristina Marconi, given to her grandson, Guglielmo, *To my dearest grandson, Guglielmo, with tender loving care, Your grandmother, 1979.*

73. Artist rendering of Marconi working with his parabolic antenna--Villa Repellini, Santa Margherita, Liguria, Italy.

74. Marchesa Maria Cristina Marconi with the Director General of Marconi Company, Mr. Paul Robinson, Mr. Peter Turrall (International Public Relations Manager), and Elettra Marconi.

75. Princess Elettra Giovanelli Marconi, Marchesa Maria Cristina Marconi and Guglielmo Giovanelli Marconi--Rome.

76. Marchesa Maria Cristina Marconi--Rome.

# THE QUEEN'S CUP REGATTA
# Kingstown 1898
# AMERICA'S CUP
# September/October 1899
# by Elettra Marconi

An exceptionally important event took place on Thursday, 21st July, 1898 in Kingstown, Ireland. It was the day of the Royal St. George Yachting Club's Regatta, called the Queen's Cup, the most famous Regatta in Ireland, and my father, Guglielmo Marconi had the brilliant idea of demonstrating the use of wireless by transmitting news of the races for the first time in the world. I remember when I was a little girl my father telling me how much he enjoyed those days when he was transmitting his messages by wireless telegraphy from the tug boat the *Flying Huntress* to the land station in Kingstown. My father loved the sea, he loved sailing himself and it was a great opportunity and a great joy for him to accomplish this first experiment at sea.

*The Daily Telegraph*, the Dublin Daily Express and the Evening Mail received the messages through wireless telegraphy. They were transmitted from the *Flying Huntress* to the land station at Kingstown and from there they were telephoned to the offices of the Express and Evening Mail. The newspapers received reports on the progress and the result of each race long before all their rivals. The greatest interest was centred in the race for the Queen's Cup and the Evening Mail was able to name the *Bona* as the probable winner long before any other paper could issue any news about the progress of the race.

Reports in the newspapers were enthusiastic: "Early in the morning the trains carried large numbers of people intent on viewing the big races dressed in their most elegant summer dresses. They were all lovers of yachting and as the sails of the yachts were hoisted every man in the big crowd had his binoculars fixed on the beautiful crafts. They were enthusiastic experts and all keen to enjoy the Royal St. George Yachting Club's Regatta, known as the Queen's Cup.

Besides the many members of the Royal St. George Yachting Club there was a large number of visitors enjoying the proverbial hospitality for which the members are renowned. The ladies took advantage of the occasion to display their newest and most fashionable gowns. The rows of spectators attired in different shades of colours gave so much variety to this outdoor display. Many of the gentlemen wore white with large white hats and the ladies the same with little umbrellas to protect themselves from the glare of the sun shining in a blue sky. Military bands played lively music, the sound of which floated seaward where the white sails of the yachts made a picture full of light and beauty. The harbour was filled with crafts of all sizes from the vast *Rainbow* with its enormous sails down to the tiny waterwags. The crowds in Kingstown increased. The wind was too light to test the speed of the big sailing yachts but there was a good breeze to allow the boats to be handled with dexterity; the delight of all yachting people."

Guglielmo Marconi had hired a tug boat, the *Flying Huntress* on which he built a very high antenna (22 ½ metres). In the cabin below were a transmitter, a receiver, an ordinary Morse tapping-machine and two batteries underneath the plain wooden table. The instruments were connected by a copper wire, attached to a wire netting which ran up to the top of the improvised mast. An observer wrote that when he stepped on board into the cabin where he was going to transmit, all those present were surprised at his appearance--"blue eyes, a resolute mouth, a boy's tall athletic figure, such is the Italian Irish inventor. His manner is confident, he speaks freely and fully and quite frankly defines his knowledge of the mysterious powers of electricity and the ether. At his instrument his face shows a suppressed enthusiasm which is a delightful revelation of his character. He explained to us that the waves for wireless telegraphy propagate at 182,000 miles per second so that no hurricane in the air has any sensible effect on them. They are also unaffected by hail, rain, snow, cloud or fog. They are unaffected by meteoric conditions and can go round large obstacles like mountains.

Marconi's invention consists in making use of electromagnetic waves or vibrations transmitting messages through the ether. A youth of twenty-three who can communicate on the wings of the wind, Signor Marconi listens to the crack-crack of his instrument with the result that the vibrations induced at one end are exactly taken up the other end by means of the well-known Morse code. All our party who were on board the *Flying Huntress* were greatly struck by the young inventor and his personal charm. There was something very striking in Signor Marconi's attitude as he stood with his hand upon the key sending waves after waves of electric oscillation in ordered succession across the Bay of Dublin. He stood there with a certain simple dignity, a quiet pride in his own control of powerful forces which suggested a great musician conducting the performance of a masterpiece of his own composing. When the first message to the shore was concluded Signor Marconi turned to the receiver instrument. We all bent forward eagerly as the strip of paper began to unroll from the recorder--the dots and dashes began to appear upon it. 'There is the answer' said Signor Marconi quietly. This great achievement, this wireless telegraphy will produce a revolution in the conditions of life; Signor Marconi's invention will bring extraordinary changes."

"It was a coincidence that for the first time Signor Marconi was showing how he has conquered space--in Irish waters--and for the first time an Italian boat the *Bona* should win the Queen's Cup and that news of it should be made known to the public by the skill of the great Italian inventor who has revolutionized the electrical world. At the same time in the quiet room of the harbour master's residence the news of the movements of the boats at sea was received without a hitch. There was a haze at sea so that those on shore could not follow what the yachts were doing but, thanks to the wireless telegraphy, each change in position and each change in tactics was sent to the land station of the Daily Express in Kingstown without delay. The Kingstown

Regatta of 1898 is assuredly the most memorable one that has ever been held."--By wireless telegraphy--*Flying Huntress.*

"At 10 o'clock sharp the steam tug especially chartered steamed off Kingstown Wharf to follow the Regatta from offshore with Signor Marconi, the Editor of the Daily Express and a number of scientific and commercial gentlemen. The yachting expert of the Daily Express stood on the bridge and gave the news of the races while Signor Marconi was operating the transmitter in the cabin of the tug boat. Thence the messages were transmitted by wireless telegraphy to the land station, whence they were immediately forwarded by telephone to the Daily Express. I could hear from the deck the crick-crack of Signor Marconi's instrument. It was a great success."

"In a room in the Harbour Master's house were the duplicates of the instruments on board the *Flying Huntress*. There was Signor Marconi's chief assistant Mr. Kemp and Mr. Glenville, a graduate at Trinity. Mr. Kemp, who has a varied experience is one of the most patient members of the electrical race. He is an old navy man. Mr. Kemp said: 'The one thing to do if you expect to find out anything about electricity is to work, for you can do nothing with theories. Signor Marconi's discoveries prove that the professors are all wrong and now they will have to go and burn their books.' Mr. Kemp remarked that Signor Marconi is a most constant and hard worker: 'He works in all weather, I remember him having to make three attempts to get out past the Needles in a gale before he succeeded. He does not care for storm or rain but keeps pegging away in the most persistent manner. Signor Marconi has been but a couple of years in England and see how much he has accomplished! The time will come, I believe, when the system will be universal'.

"What are the practical possibilities of the invention? For communicating with ships at sea--all light house stations will be equipped with transmitting and receiving apparatus. In case of fog the possibility of being able to communicate with land will greatly lessen the danger of accident. We may expect the fleet to be equipped with the required instruments. For journalistic purposes there are many circumstances in which the invention will prove of the highest value. The occasion was one which illustrated the application of Signor Marconi's wonderful invention to the purpose of journalism."

"The second day's experience of reporting by Wireless Telegraphy was even more interesting than the first. The weather was thicker so that the distant parts of the yacht races could not be observed from the shore. The boats making for the Kish lighthouse disappeared from the sight of the shore observers. The only witness of their movements was the *Flying Huntress* and through Signor Marconi's operating he was able to communicate with Kingstown. The bulletins of the receiving station were the only means to learn of the progress of the race. The half past one edition of the Evening Mail which was sold in Kingstown by 2 o'clock contained these dispatches: There was a very heavy sea and Signor Marconi's instruments were installed in a makeshift fashion. The salt spray which broke constantly over the tug was drenching every now and then Signor Marconi's induction coil in the cabin. The instrument was seriously interfered with. But the fact that it triumphed over such difficulties was a more striking proof of the success of the invention than any test it had yet sustained. For Wireless Telegraphy it was an even more satisfactory day than the preceding one. The messages were transmitted by Wireless Telegraphy to the land station. The greatest distance of communication that has been reached is 14 miles and Signor Marconi states that foggy or stormy weather rather facilitates the passage of the electric waves. This is the result of today's experiments: The messages, however long, were transmitted with absolute accuracy and did not require to be repeated in a single instance. The distance from which they were sent varied from five and ten miles. The last edition of the Evening Mail contained two full columns of yachting".

The Director of the New York Herald was very impressed by the news of the successful radio reporting of the Kingstown Regatta and he invited my father to report on the America's Cup which was to be held in September 1899. The Marconi Company was enthusiastic at the thought of the publicity but Marconi wanted to concentrate on his work for the first transmission across the Channel from France to England. He agreed that if all went well with the cross-channel transmission he would leave for America in September.

The transmission across the Channel in March 1899 was a complete success and was widely reported in the American press. Marconi arrive in New York aboard the *Urania* on 21st September 1899. He was met by a crowd of reporters all anxious to see the famous inventor. They were surprised to find a quiet unassuming young man with fair hair and bright blue eyes--nothing like the stereotype Italian they had expected to see. They were even more surprised when they found that he spoke fluent English without a trace of a foreign accent. A reporter from the Herald wrote: "Very few of the numerous people waiting on the quayside recognized in that young man, who seemed almost a boy, the man whose name has become famous in the scientific sphere". Another reporter wrote: "No taller than a Frenchman and no older than a quarter of a century. He is only a boy with a boy's cheerful character and enthusiasms but he seems to have the dynamic conception of life meaning work that is typical of an adult". Yet another described him as "a serious and somewhat reserved young man who speaks little but always to the point". Marconi declared: "We will be able to transmit the details of the yacht races to New York with the same precision and the same speed as a telephone call. The distance does not count nor will the mountains obstruct the transmission.

The American yacht was J.P. Morgan's *Columbia II* while the British challenger was Sir Thomas Lipton's *Shamrock*. Marconi set up his transmitting equipment first on a ship called the *Ponce* and later on the *Grand Duchesse* in which he followed the *Columbia* and the *Shamrock* during their races. At the start of the regattas he transmitted a message declaring that thanks to the wireless telegraph the races taking place offshore from Sandy Hook were only a minute away from Herald Square. The radio apparatus worked perfectly and Marconi's messages transmitting the details of the races reached the Herald in New York at a rate of 15 words per minute, a great speed for those days.

The final stop-press news was that the *Columbia II* had won the America's Cup. In a leading article the Editor of the Herald wrote: "The possibilities offered by the application of the

Wireless are so important that not only he who takes an interest in science but also anyone who wants to send a telegram must have at heart every initiative aimed at making the system known to the public and demonstrating what it is capable of achieving from the commercial point of view".

# THE FIRST TRANSMISSION ACROSS THE CHANNEL
# by Elettra Marconi

On 27th March, 1899 Marconi made the first transmission by wireless telegraphy between France and England from Wimereux near Boulogne to the South Foreland lighthouse near Dover, a distance of 32 miles across the Channel.

At the beginning of March he had received the official approval of the French Government to set up a wireless station at the Chalet d'Artois at Wimereux three miles from Boulogne sur Mer where his parents had married thirty-five years before. French visitors to the station were surprised to see the young inventor himself hard at work putting up the mast. Marconi always liked to work personally on every new wireless station to make sure that everything would work perfectly. His assistants admired him for this because they knew that he was working harder than everyone else and they were inspired and encouraged by his example.

On 27th March the French officials and dignitaries arrived at Wimereux to witness the historic transmission.There were delegates from the War Ministry, the Navy and the Ministry of Trade. At 5 o'clock in the afternoon Marconi pressed the transmission key in the wireless room. He concluded his message with the agreed signal: VVV. Then he leaned back quite calmly waiting for the answer. He knew that the experiment would work. A moment later the reply from the South Foreland lighthouse began to appear on the tape: "V M" (Your message perfect). Marconi transmitted: "The same here. 2CMS.VVV" (Victory)! That first transmission made by electric waves across the channel was a sensational event. The Times of 28th March, 1899 published a description of wireless telegraphy by the newspaper's foreign correspondent in Boulogne--transmitted by wireless across the Channel! To prove that the transmission worked perfectly in both directions Marconi took the Commission across the Channel to South Foreland and messages were sent back and forth across the Channel without the slightest hitch.

On 30th March, in a supplementary note to a conference he had given at the Italian Royal Academy at the beginning of March, Marconi wrote: "...communications between France and England were established for the first time on 27th March. On the English coast the station is at the South Foreland lighthouse near Dover and on the French coast at the Chalet d'Artois at Wimereux near Boulogne. On the previous Monday the apparatus was sent from London, entrusted to two of my assistants. A house was leased in which to set up the station. A mast was then erected and at 5 o'clock on 27th March, one week after the instruments had been sent from London, perfect telegraphic communications were established between the two points. The first transmissions were carried out in the presence of a Commission, nominated by the French Government, made up of Colonel Conte du Pontavice de Heussey, Captain Ferrié and Monsieur Voisenet. The first messages were sent from France to England and received an immediate reply from the operator in charge of the station at the South Foreland lighthouse. Since then there has not been the smallest hitch in the communications and you will be interested to hear that yesterday, 29th March, the service was carried out by two French officers: Captain Ferrié, a French engineer at the English station and Monsieur Voisinet, a French telegraphic engineer at the French one. These gentlemen kept up telegraphic communications for several hours and then, together with many other people, offered their congratulations on the excellent operation of the apparatus".

A few days after the first transmission across the Channel, Cleveland Moffet, a journalist for the American McClure's Magazine, went to Wimereux while Robert McClure, the publisher of the periodical went to the South Foreland lighthouse. They wanted to prove to themselves that wireless was doing what people were saying, actually transmitting without wires. The message was given to Mr. Kemp at Wimereux to transmit to England with the words spelt backwards: McClure, Dover: "gniteerg morf ecnarf ot dnalgne hguorht eht rehte--Moffet. Back came the answer: Moffet, Boulogne: "Your message received, it reads all right. Vive Marconi--McClure". From then on McClure wrote in his magazine that Marconi was a brilliant young man who could be trusted and admired. He wrote a series of articles describing Marconi's progress and he did much to bring Marconi into the public eye both on the Continent and across the Atlantic Ocean.

Marconi sent many more messages successfully across the Channel. The weather was very bad all through the month of April but the storms did not affect the transmissions. Marconi was continuously on board the warship Ibis and the cargo boat Vienne, building his wireless apparatus, transmitting and receiving so that the men of the French Navy could learn from him how to make his wireless work. On 17th June, George Kemp, Marconi's trusted assistant, went to welcome him when he disembarked from the Vienne. During the carriage drive from Boulogne to the Chalet d'Artois the horse suddenly bolted and the carriage was upset. Marconi was thrown out and injured his knee. He had to stay immobile for the next week but fortunately his knee recovered quickly. He enjoyed playing the piano and in the evenings he and his assistants would relax after a hard day's work while he played their favourite music and they all sang the fashionable songs of the day.

He received many important visitors in his radio station in Wimereux. They all wanted to see the transmitter and the receiver and understand how the messages were sent and how Marconi could receive the messages from England. One day Lord Baden-Powell, the founder of the Boy Scouts, visited him. Marconi sent a message to South Foreland but no matter how hard he tried he could get no answer. He checked all his connections, inspected the wires leading to the aerial, changed the coherer but still no answer came. Baden-Powell started to feel worried and embarrassed. Perhaps wireless telegraphy was not so reliable as it had seemed. All of a sudden they heard the coherer was clicking again and the tape started to unroll from the Morse printer. They both read: "Just back from supper. Anything happening your end?"

Marconi was always generous and ready to give credit where credit was due. One of his first messages from South Foreland was to Professor Branly, the great French scientist: "Marconi sends M. Branly his respectful compliments across the Channel. His fine achievement being partly due to the remarkable researches of M. Branly". In the *Bulletin de la Société francaise de Physique* ("Résumé des Comunications". Seance du 16 décembre 1896, pag. 78 del volume del 1898.) Professor Branly had written: "Although the experiment that I have always presented as the most important of those made during my studies on radioconductors (some batteries, a tube with iron filings and a galvanometer, comprising a circuit in which the current flows after an electric spark has been struck in the distance) corresponds to the concept of the wireless telegraph, I do not claim to have made this discovery because I never thought of transmitting signals".

In 1905, in an article in the French magazine *Je sais tout* Professor Branly wrote: "Wireless telegraphy is something totally strange, totally scientific, totally simple. This is why we had to wait until the year 1896 to begin to understand it. In fact the simple solutions to major problems are always the most difficult to find. For years scientists go around the problem and get close to the solution; then, one fine day in the midst of all their research a ray of light appears: philosophic deduction has dissipated the mist that hid progress. The solution appears bright and clear! Thus for some time great scientists had been studying the propagation of electric vibrations. One of them, Professor Hertz has linked his name to them, calling them waves. Waves! That is to say the sudden change in vibrational equilibrium produced by any electrical discharge. It is easy to produce these waves and Professor Hertz published some notes on this subject which were of interest only, or mainly, to academics. A young scientist, Mr. Marconi arrived who thought that since it was possible to produce electric waves and to project them into space, it might also be possible to pick them up at a distance and "converse" as the specialists say. Mr Marconi had the merit of rapidly inventing some ingenious equipment to receive the waves and this in spite of the doubts and opposition which his courageous idea had to contend with. Wireless telegraphy was born. We will not go into the whole story here. Universal Science has agreed to develop this wonderful discovery since the young Italian scientist lifted the veil behind which its practical use was hidden.

In 1937, immediately after Marconi's death, Professor Branly was interviewed by a journalist of Le Figaro in his Parisian laboratory. Branly declared: "The death of my famous brother of the Pontifical Academy of Sciences has deeply affected me. He was a scientist, an honest, noble and generous man. Although I only met him once I have a vivid memory of him. In 1912, in this same laboratory, he came to offer me the post of director of technical sciences of the company of which he was the Director. I refused. Our lives were different. A man of science but an expert engineer Marconi created an industry from laboratory work. He was a man of scientific accomplishment. I was only interested in scientific research. I have lived in my laboratory for 62 years. It has been my life. Wireless telegraphy from the very beginning had passed into a sphere which was not mine. I was already searching for something else. But Marconi's generous heart, his scientific honesty in acknowledging the role of those who had gone before him, his research guided by a brilliant intelligence deserve the highest praise. His country is not alone in its loss".

# THE ARREST OF *Dr* CRIPPEN
# by Elettra Marconi

I always remember a story that struck my imagination when I was a little girl: the first arrest accomplished by radio; that of the notorious murderer "Dr" Crippen. He had killed his wife and hidden her body under the floor of his house. He and his girlfriend Miss Le Neve travelled to Holland and sailed aboard the *Montrose* for Canada. They had false passports and the girl was dressed as a boy. They travelled as Mr. Robinson and son. The police had issued a warrant for their arrest and a description of the fugitives was published in all the newspapers.

Not long after the ship left Antwerp the Captain noticed that there was something strange about the couple: the boy's clothes did not fit him very well and he seemed uncomfortable in them and the father held his son's hand rather too affectionately. His suspicions aroused, Captain Kendall realized that they looked very like the missing pair. He had been aboard the first British ocean-going ship to be equipped with Marconi's wireless system and he immediately realized the important role that wireless could play in bringing a criminal to justice. On July 22$^{nd}$ 1910 he sent a message by wireless to Scotland Yard.

The next day, Chief Inspector Dew sailed aboard the *Laurentic* which was a faster ship than the *Montrose.* While the couple were travelling across the Atlantic, completely unaware that the chase was on, news of them was in every newspaper and the chase across the Atlantic was the main topic of conversation. When the *Laurentic* overhauled the *Montrose* Inspector Dew went on board the *Montrose* with two other police officers, disguised as pilots. Dr Crippen was walking on deck and once Inspector Dew had taken a good look at him he gave the signal to arrest him together with Miss Le Neve.

# THE RESCUE OF THE TITANIC by Elettra Marconi

In November, 1996 I went to Southampton for the 100th Birthday of Mrs. Edith Hayman, one of the last living survivors of the disaster of the *Titanic.* I also met a French gentleman and two other ladies who were survivors of the shipwreck. I had already met them in September at a reception near the Hudson River in New York after their return from a voyage (organized by Mr. George Tullock) to visit the place where the *Titanic* sank. After hearing their stories I visited the Museum in Southampton where I listened to the recorded voices of many survivors who had been interviewed after the tragedy, telling about their experiences and how they had been saved.

Now, when the story of the *Titanic* is so much in the news, my thoughts go back to when I was a little girl. I remember hearing my father speak about the sinking of the *Titanic.* I shall never forget the emotion in his voice when he spoke about the tragedy. Who knows how many more of the passengers and crew could have been saved if only the S.O.S calls sent out by the *Titanic's* wireless operators had been heard by the *Californian,* the only ship close enough to come to their rescue in time. As it was, although over seven hundred passengers were saved thanks to the radio invented by Guglielmo Marconi, one thousand five hundred people lost their lives in the freezing sea. He always spoke with admiration of the heroism of those two young "Marconi men", Jack Phillips and Harold Bride, who continued to transmit the distress call until just a few minutes before the ship sank. My father, who was hailed as the saviour of the

*Titanic,* always gave the greatest credit to those brave men. Jack Phillips' heroism and devotion to duty cost him his life.

The GEC-Marconi Ltd.have given me the opportunity to consult the original radio messages received from the *Titanic* before and during the disaster. I have included this precious new material in the following account of the role of the radio in the rescue of the *Titanic.*

Ever since boyhood Marconi, who was himself a sailor, had been acutely aware of the isolation of ships at sea. Now, thanks to his invention, ships sailing the oceans hundreds or even thousands of miles from land and hundreds of miles from each other could send and receive messages. Three years before the *Titanic* disaster he had been acclaimed as the benefactor of humanity because his invention had made it possible to rescue another transatlantic liner, the *Republic.* Just after leaving New York bound for Europe it came into collision with a second ship, the *Florida,* in a dense fog some 26 miles southwest of the Nantucket Lightship. Fortunately the *Republic* was equipped with a Marconi wireless system. The C.Q.D. distress signal transmitted by the wireless operator on board the *Republic* was received both on land and by other ships so that almost all those on board survived the disaster.

The *RMS Titanic,* the pride of the White Star Line, built in Belfast, Ireland, left Southampton on her maiden voyage on Wednesday, 10th April, 1912. She had been designed to be unsinkable and was referred to as a "floating castle". Some wealthy American passengers paid $5,000 for a first-class suite (It would have taken the senior Marconi wireless operator18 years to earn this amount) while steerage passengers paid £7.75p.The *Titanic* sailed to Cherbourg in France and then to Queenstown in County Cork, Ireland where she took on more passengers. Many ships sent wireless messages of congratulations and good wishes to the *Titanic* as she steamed towards New York.

The night of Sunday,14th April, 1912 though moonless was starry, the atmosphere exceptionally clear and the sea absolutely calm. In the dark hours between sunset on Sunday and dawn on Monday, 15th April the *Titanic* met an ice field which had floated down from the Arctic sea.

It was unusual for ice to be found in the North Atlantic Ocean at this time of year but other Atlantic steamers had passed through the same ice field and several wireless messages warning of icebergs had been received by the *Titanic*. On Saturday,13th April the *Caronia* had sent out a message stating: "Westbound steamers report bergs, growlers and field ice latitude 42 N. from longitude 49 to 51 W". On Sunday morning a message was sent from the *Amerika* to the Hydrographic Office in Washington D.C. via the *Titanic* and Cape Race: "*Amerika* passed two large icebergs in latitude 41.27 N. longitude 50.08 W on the 14th of April". Another message was sent on Sunday morning to Captain Smith of the *Titanic* from the Captain of the *Baltic* saying: "Have had moderate variable winds and clear fine weather since leaving. Greek steamer *Athinai* reports passing icebergs and large quantity of field ice today in latitude 41.51 N. longitude 49.52 W... Wish you and *Titanic* all success". Captain Smith replied: "Thanks for your message and good wishes. Had fine weather since leaving". The *Caronia* transmitted another ice warning from the *Noordam* to the *Titanic* on Sunday afternoon: "To Captain *S.S. Titanic*. Congratulations on new command. Had moderate westerly winds fair weather no fog much ice reported in latitude 42.24 N to 42.45 N and longitude 49.50 W to 50.20 W". The most critical report was received from the *Mesaba* on Sunday evening: "To Titanic and all East Bound Ships. Ice reports in latitude 42N to 41.25 N longitude 49 W to longitude 50.30 W. Saw much heavy pack ice and great number large icebergs also field ice. Weather good clear. Reply: Received, thanks". Amazingly, Captain Smith and his First Officer did not act on the ice warnings apart from posting lookouts in the crow's nest to watch for icebergs and the *Titanic* continued to steam on at high speed. At 11.40 p.m. the lookouts shouted the report of "Iceberg dead ahead!" to the bridge. The *Titanic's* position was then 41.46N. 50.14 W, exactly where the ice field had been reported. Less than a minute later the *Titanic* struck the iceberg, ripping open the hull for a length of nearly 100 metres. At 2.20 a.m. the invincible *Titanic* split in two and sank.

Another passenger ship, the *Californian* was only ten miles away when the Titanic struck the iceberg. Cyril Evans, the wireless operator of the *Californian,* giving evidence at the British Inquiry into the *Titanic* disaster in May, 1912 said that on Sunday afternoon the *Californian* sent a message to the Captain of the *Antillian* saying: "Latitude 42.3 N. Longitude 49.9 W. Three large bergs five miles to Southward of us". A little later he made contact with the *Titanic* and gave the same report about the icebergs and received the reply: "All right, I heard the same thing from the *Antillian*". At 11 p.m. ship's time the captain ordered him to tell the *Titanic* that the *Californian* was stopped and surrounded by ice. Evans sent the message to the *Titanic* and got the reply "Keep out" because the *Titanic* was at that moment in communication with the Cape Race receiving station near St. John's in Newfoundland and his message had interfered with the *Titanic's* transmission. Being the sole wireless operator and having put in a long day, Evans retired for the night. Early the next morning the First Officer came into his cabin and said "There's a ship been firing rockets, will you please try to find out whether there is anything the matter?" He immediately jumped out of his bunk and took up the telephone but could hear nothing. He then sent out a general call C.Q. and got an answer from the *Mount Temple* saying: "Do you know the Titanic struck an iceberg and is sunk?" That was the first the *Californian* knew of the disaster and the desperate distress calls transmitted by the *Titanic's* wireless operators went unheard by the only ship close enough to reach the sinking *Titanic* before it was too late.

Jack Phillips, the chief Marconi man on the *Titanic* was a native of Godalming in Surrey, England. He had started his career as a wireless operator on ships but had then been posted to the Marconi Company transatlantic wireless station near Clifden in Co. Galway on the west coast of Ireland. After three years at Clifden he applied for a return to operating at sea. In the spring of 1912 he was appointed chief wireless operator on the *Titanic.* He celebrated his twenty-fifth birthday on April 11th, 1912. The assistant wireless operator on the *Titanic* was Harold Bride from Nunhead in England. The two wireless operators had joined the ship at Belfast at the beginning of April and on the trial trip of the ship from Belfast to Southampton they had tested the wireless apparatus and found it in good working order. The installation on the *Titanic* was a very modern type and the most powerful of any ship of the merchant navy at the time. It was guaranteed for a distance of about 350 miles although in actual practice it carried a great deal further. It utilised a wireless telegraphy spark transmitter (using signals in Morse Code) while the receiving equipment included the famous Marconi magnetic detector. The apparatus was in duplicate and there was a spare battery so that it could be operated in case the current from the dynamos was cut off owing to the engines being flooded.

1900 saw the birth of the Marconi International Marine Communication Company Ltd. Since that year, when the first large liner was fitted with a wireless system, the Marconi Company had provided hundreds of ships with wireless systems. All wireless stations on ships or on land had identifying "call signs" beginning with the prefix 'M'. The call sign of the *Titanic* was MGY. The operators on most of the ships were Marconi Company employees and they wore caps with an 'M' embroidered on the front.

Marconi himself gave evidence during the official Inquiries into the Titanic disaster in the United States and Great Britain. He offered spontaneously to testify before the United States Senate Committee which was investigating the causes of the wreck of the Titanic. During the Inquiry he was asked: "Who was the first practical operator of wireless telegraphy covering long distances" and his answer was: "I think it was myself, in England in 1896 and 1897... I took an interest in electrical subjects generally. I had studied a great deal. I was what I might rightly describe as an amateur". He was asked to make a brief statement describing his work and he answered: "I first carried out some tests in Italy with electrical waves... I invented apparatus which made them apparent or made it possible to detect them over 2 or 3 miles. That was at the time considered very interesting. After that I came to England where I had numerous relations and I offered to demonstrate this new idea to the British post office, the army and the navy and to Lloyd's. They were very greatly interested in the system and tests were carried out and communication was very shortly established over 9 miles. The first British ship that was fitted was a yacht belonging to the late King

Edward and several warships belonging to the British Navy and the Italian Navy.

The system worked very well up to a limited distance. It was nowhere near as reliable as it is now. After a certain space of time, in 1899 and 1900, some further improvements were perfected by myself and some by others which greatly increased the range and made it apparent at once that it would be possible to communicate over thousands of miles and steps were taken for the installation of stations to carry out tests to show if it were possible.

The first tests in America were carried out by myself, in 1899, at which time I also carried out experiments on battleships of the United States Navy, the *New York* and the *Massachusetts*. Communication was established, I think, up to 20 or 25 miles, or something like that, at that time... At present the useful reliable range is something like 3,000 miles... I expect it will be one of the principal means or methods for communicating between distant parts of the world...for communication, say, between New York and England, or between New York and San Francisco, or between Chicago and another distant place, I think that with the increase of speed and the understanding of electricity it will some day become the chief means of communication".

Marconi was asked about the rules and regulations governing wireless operators on board ship. He told the Committee that they were employees of the Marconi Company but took day to day instructions from the ship's captain. A large ship like the *Titanic* always carried two operators so that a continuous service was maintained. One of the operators always had the telephone fixed to his ears and could hear any call which was made although he could talk or read when he was not actually receiving or sending a message. Smaller ships like the *Carpathia* carried only one wireless operator. The *Carpathia* carried a short-distance wireless equipment; an apparatus which could transmit messages, under favourable circumstances, up to about 180 or 200 miles, but on average about 100 miles depending on the state of space at the time and to a large extent on the skill of the operator. The *Titanic* was equipped with a more powerful wireless apparatus capable of communicating with accuracy over 400 or 500 miles during the daytime and very often 1,000 miles during the nighttime. He thought all ships at sea should carry two wireless operators so that an operator was constantly at his key.

At the British Inquiry, in answer to a question on whether it would be possible, on a ship which was manned by one operator, for a person who was not an expert in wireless telegraphy to receive some simple signal so that he could then call the operator, Marconi replied: "I have another way that suggests itself to which I have given a great deal of attention since the *Titanic* disaster and that is of making the wireless apparatus ring a bell and thereby give warning that a ship in danger needs assistance...some tests have been made with an apparatus such as I have referred to and I have considerable confidence that it can be employed".

Giving evidence at the British Inquiry Harold Bride told the court that he and Jack Phillips had agreed that Phillips would go on duty from 8 o'clock at night until two in the morning and Bride from 2 o'clock in the morning until eight. During the day they took turns to suit each other's convenience but a continuous and constant watch was kept and one or other of them was always in the Marconi room, close to the bridge. At around 5 o'clock in the afternoon (ship's time) on Sunday 14th April he received the message warning of icebergs from the *Californian* and immediately delivered it to the officer on the bridge. After dinner he went to bed but relieved Phillips at 12 o'clock midnight which was two hours earlier than usual as Phillips had been very busy the night before. This was after the impact with the iceberg. He had been asleep at the moment of the collision and the first he knew of it was when Phillips told him he thought the ship had struck something from the feel of the shock that followed. In the words of Harold Bride: "I was standing by Phillips telling him to go to bed when the captain put his head into the cabin. 'We have struck an iceberg', he said; 'you had better get ready to send out a call for assistance. Don't send it until I tell you'. The captain went away and in ten minutes he came back. We could hear terrible confusion outside but not the least thing to indicate any trouble. The wireless was working perfectly. 'Send a call for assistance', ordered the captain. 'What call shall I send?' Phillips asked. 'The regulation international call for help--just that', was the reply; and Phillips began to send the signal C.Q.D., joking while doing so".(This detail in Bride's testimony shows that Phillips was facing this terrifying moment with the greatest courage--E.M.). "After a few minutes however, the captain reappeared and said, 'Send S.O.S.; it may be your last chance'.The *Carpathia* answered our signal and we told her our position and said we were sinking by the head. The operator went to tell the captain and in a few minutes returned and told us that the *Carpathia* was putting about and heading for us". This was the first emergency use of the distress signal S.O.S. which had taken the place of the old distress call C.Q.D. In fact, in 1906 the International Radio-telegraphic Convention had laid down principles and regulations governing wireless telegraphy at sea and at that time the distress call was altered to S.O.S. although C.Q.D. which was so well known continued to be used as well.

Fifty-eight miles away from the *Titanic,* Thomas Cottam the wireless operator of the steamship "*Carpathia*" had been preparing to retire for the night when he received the distress call from the *Titanic*: "CQD. Struck iceberg. Come to our position. 41.46N. 50.14W." The ship's log on board the "Carpathia" in fact shows the annotation: "Heard the *Titanic* invoke CQD and SOS", and ten minutes later: "Change course". The "*Carpathia*" altered its course and steamed to the rescue, guided by the radio signals transmitted by the Titanic's heroic wireless operators who remained at their posts in the wireless station until minutes before the *Titanic* sank. Continuing his evidence during the British Inquiry, Harold Bride said that after Phillips sent the C.Q.D. sign they received answers from the *Frankfurt* and the *Carpathia* while they received several messages from the *Olympic* right up to the time when they finally left the wireless cabin. At 10.50 p.m. (ship's time) the *Olympic's* wireless log reads: "Hear MGY (*Titanic's* call sign) signalling to some ship and saying about striking iceberg, not sure if it is MGY who has struck iceberg". Bride went to report to the captain who was on the boat deck superintending the lowering of the lifeboats. Later the captain came into the Marconi room and told them the ship would not last very long and that the engine-room was flooded. Phillips later went outside to look round and when he came back he said that the fore well-deck was awash and that they were putting the women and children in the boats and clearing off. Then the captain came in and told them to shift for themselves, because the ship was sinking. Phillips took the telephones up when the captain had gone away and started to work again. Bride could read what Phillips was sending but not what he was receiving and he judged that the *Carpathia* and the *Frankfurt* had both called up together; the *Frankfurt* was interfering with Phillips's reading of the *Carpathia's* message. Phillips told the operator of the *Frankfurt* to "keep out of it and stand by". He then told the *Carpathia* that they were abandoning the ship. Phillips tried to call once or twice more but the power was failing and they failed to get any replies. Then he and Phillips lined up on top of the Marconi cabin in the officers' quarters. They were trying to fix up a collapsible boat and he helped to get it down from the top deck to A deck. He got into it but as the ship sank it floated off upside down. He was swept off the boat deck. When he last saw Phillips he was standing on the deck-house. Bride swam away from the collapsible boat but joined it later. He was rescued by the *Carpathia* early on Monday morning. Tragically, Jack Phillips perished in the disaster.

Many other ships received the *Titanic's* distress call and altered course to go to her rescue.The *Mount Temple* was about 50 miles south west of the *Titanic*. John Durrant, the Marconi operator of the *Mount Temple,* giving evidence at the British Inquiry, told the court that on Sunday night at 11 minutes past midnight (ship's time) he got the message CQD from the *Titanic* giving her position and adding "Come at once. Struck berg. Advise captain". He told the captain at once and about 15 minutes after getting the first signal the *Mount Temple* had altered course and was speeding to the assistance of the *Titanic*. At 12.34 a.m. he heard the *Frankfurt* answer the *Titanic's* CQD call and the Titanic immediately gave her position and asked "Are you coming to our assistance?" The *Frankfurt* replied "What is the matter with you?" and the *Titanic* answered "Have struck an iceberg. Sinking. Come to our help. Tell captain". The *Frankfurt* then said, "O.K.Will tell bridge at once", and the *Titanic* replied, "O.K. Yes. Quick". At.

12.42 a.m. he heard the *Titanic* call SOS. At 12.43 a.m. he heard the *Titanic* call the *Olympic* and at 1.06 a.m. the *Olympic* replied and got the message, "Going down fast by the head", and then "We are putting the women off in the boats". At 1.29 a.m. the *Titanic* sent out a call, "CQD. Engine room flooded". At 1.33 a.m. the *Olympic* sent a message to the *Titanic* asking "Are you steering southerly to meet us?" but the only reply from the *Titanic* was the code word for "Received". That was the last message the operator of the *Mount Temple* heard from the *Titanic*. The *Olympic*, the *Frankfurt*, the *Birma* and the *Baltic* were all speeding to the rescue and continued to call the *Titanic* but there was no reply although the operator of the *Virginia* thought he heard a faint C.Q. call at 2.27 a.m. The operator of the *Mount Temple* at 2.36 a.m. made the entry "All quiet now. The *Titanic* has not spoken since 1.33 a.m.". When the messages ceased he thought the flooding of the engine room had put the wireless out of condition. Most ships carried storage batteries for use when power could not be obtained from the dynamos and the wireless apparatus could be changed from the dynamos to the storage batteries in a minute; but the range of a wireless using storage batteries would be less than that of a wireless using dynamos.

At 4.46 a.m. he made the entry, "All quiet. We are stopped away. Pack Ice". At 5.11 a.m. the *Californian* called C.Q. (the call to all stations) and he answered, telling her that the *Titanic* had struck an iceberg and sunk. His last entry was "8 a.m. Heard from *Carpathia* that she had rescued 20 boatloads". The *Carpathia* was the first to arrive on the scene at 4.15 a.m. on Monday to find only the lifeboats containing seven hundred and five passengers. Three hundred and twenty-eight bodies were recovered from the sea. The position of the *Titanic* had been given at 41.46N 50.14W but the *Carpathia* found the survivors at 41.43N. 49.56W. approximately 13.5 miles east-southeast of this position. A message from the *Carpathia* to the *Olympic* on 15th April said: "South point pack ice in 41.16 N. Don't attempt to go North until 49.30W. Many bergs large and small amongst pack also for many miles to Eastward. Fear absolutely no hope searching *Titanic's* position. Left Leyland *S.S. Californian* searching round. All boats accounted for. About 675 souls saved. Latter nearly all women and children. *Titanic* foundered about 2.20 a.m., 5.47 G.M.T. in 41.16 N. 50.14 W. Not certain of having got through. Please forward to White Star also to Cunard Liverpool and New York and that I am returning New York. Consider this most advisable for many considerations.--Rostron."

Marconi himself had been invited to sail to New York on the *Titanic's* maiden voyage but he was in a hurry to get to America as he had a great deal of work to do there. He cancelled his booking and travelled from England on the transantlantic liner *Lusitania* which was due to arrive in New York before the *Titanic*. He had just arrived when the news of the disaster was received. When the *Carpathia* reached New York he immediately rushed on board to speak to the two Marconi men, Thomas Cottam of the *Carpathia* and Harold Bride of the *Titanic*. When Marconi returned on shore he made the following declaration: "I am eternally thankful that over seven hundred persons have been saved by wireless although I know that others should not have died. It is worth having lived so that these people could be saved... I know you will understand me when I say that all those who have worked with me are sincerely grateful that the wireless has once again made it possible to save human lives".

My mother often repeated the story of the scene on the docks. She said: "When the passengers saw that the inventor of the wireless, Guglielmo Marconi was standing there, calm and smiling, there were really touching scenes of emotion and enthusiasm. The gratitude of both the survivors and their relatives waiting for them on the quayside was indescribable. Everyone was crying and trying to kiss and embrace him. They even pulled all the buttons off his coat as keepsakes!" Some time later the survivors of the *Titanic* presented him with a magnificent commemorative gold medal as a sign of their gratitude. Engraved upon it is a picture of the shipwreck and the survivors with their arms outstretched towards Marconi, calling for help with the SOS signal. My mother always treasured this medal as I do to this day.

The following extracts from the British Press express the nation's gratitude to Guglielmo Marconi:

"We owe it to patient research in a delicate and difficult branch of science that the *Titanic* was able, with wonderful promptitude, to make known her distress and to summon assistance. But for wireless telegraphy the disaster might have assumed proportions which at present we cannot measure; and we should have known nothing of this occurrence for an indefinite period. Many a well-found ship has, in fact, disappeared in these berg-haunted waters without leaving a sign to indicate her fate. Thanks to Marconi's apparatus, it is now hardly possible for any vessel equipped with even moderately powerful instruments to be lost on any frequented route without being able to communicate information and to summon help. The *Titanic* had the call upon a circle of at least three hundred miles radius even in daylight,while at night the range of her instruments would be doubled or trebled. She could speak to the shore and to every vessel over that enormous area of ocean and she could be spoken to and assured that help was on the way. Not only so, but the ships appealed to could communicate with one another, act in concert, and transmit the news to indefinite distances. The advantages conferred by this abridgment of space are enormous. No vessel need be alone, none need vanish without a sign from human ken, and in none but crushing and instant disasters need any despair of help. This is surely one of the greatest of the many boons conferred upon humanity by patient, persistent and often very discouraging inquiry into natural laws...few besides experts have the faintest conception of the difficulties to be overcome, or of the mental and moral equipment needed to overcome them, when the hints are few and obscure, when every instrument has to be called out of the void, and when hope of gain, if considered at all, was infinitely remote".--*The Times*. April 16th.

"The imagination is struck once more by the wonderful part played by wireless telegraphy in the story of the *Titanic*. The wounded monster's cry of distress sounded through the latitudes and longitudes of the Atlantic, and from all sides her sisters, great and small, hastened to her succour. But for this new instrument of communication it might have been that the greatest product of naval architecture would have passed from our human ken, her fate ever unknown, or unknown at least until one of more of her boats struggled to the Newfoundland shore...The wonder of the wireless is once more demonstrated. We recognise, with a sense near to awe, that we have been almost witnesses of the great ship's death-agony".--*Pall Mall Gazette.* April 16th.

"With this means of communication (wireless) the terrible isolation of mid-ocean has vanished for ever. Her appeal for aid was received by half a score of ships and taken in by the nearest land station. From the moment when it was made her passengers and crew had the comforting knowledge that help was coming up from all quarters. Every ship within range hurried to her assistance, but it was impossible to avert loss of life".--*Daily Mail.* April 16th.

"But for the wireless what would have been the state of the unfortunate people wrecked? They might have drifted about for days looking in vain for the help that did not come, and there might have had to be told over again the story of privation and death with which the history of the sea has made us only too familiar."--*Portobello Advertiser,* April 20th.

"Never before has the romance of wireless been brought so vividly to the imagination of two hemispheres as by the news reporting the disaster of the *Titanic.* Who could fail to have been thrilled by the brief word pictures of the *Carpathia,* the *Virginian,* the *Olympic,* the *Baltic,* and other great transatlantic liners speeding hundreds of miles across the waste of waters to their sister ship in her hour of need?"--*Manchester Weekly Times,* April 20th.

"Even while the great liner was reeling back from the shock of the fearful impact the Marconi operators were at their places, and those poignant appeals for help--mute, invisible--were flying outwards on their instantaneous errand. The *Virginian,* steaming through the darkness 170 miles away, noted the call and instantly turned to the rescue. The *Olympic* picked it up, and the bells rang, too, in the telephone-room of the *Baltic* 200 miles below the horizon... There is a new sense of the value of the wonderful invention which was able to summon aid when aid could have been obtained in no other way".--*Daily Telegraph,* April 16th.

On 18th April. 1912 the Right Honourable Herbert Samuel, M.P., the British Post- Master General in a speech at the dinner of the London Chamber of Commerce said: "Those who have been saved have been saved through one man, Mr. Marconi, whose wonderful invention is proving not only of infinite social and commercial value but of the highest humanitarian value as well."

For my father, the knowledge that his invention had made it possible to save so many lives was the greatest reward he could receive for all his work but he was still not satisfied because many more lives could have been saved. He was determined to fight for new rules to be made governing wireless services aboard ship. The most significant result of his efforts was an International Radio-Telegraphic Convention which convened in London on 5th July, 1912 to establish regulations and procedures governing wireless services aboard ships and ship-to-shore. It was attended by sixty-five countries and new regulations and procedures were enacted. Marconi continued his efforts and the first of a series of conferences, "Safety of Life at Sea" was held in London in November, 1913. Sweeping regulations were put into effect governing all ships at sea. All ocean-going passenger ships were obliged to be fitted with a wireless installation and furthermore the wireless station was to be manned twenty-four hours a day. The wireless room became the foremost station on board the ship, establishing safety as the first priority. Ships equipped with wireless were sent out to patrol the North Atlantic shipping lanes to report the position of icebergs. The value of wireless on board ocean-going ships was now evident.

I should like to conclude this chapter with the words of the famous cartoon in Punch Magazine after the saving of the *Titanic*: "SOS" (Punch to Mr. Marconi) "Many hearts Bless you today Sir. The world's debt to you grows fast". How true it was then and how true it is today!

Phothos presented
for the first time
by
ELETTRA MARCONI

1. Palazzo Marescalchi where Guglielmo Marconi was born on April 25, 1874.

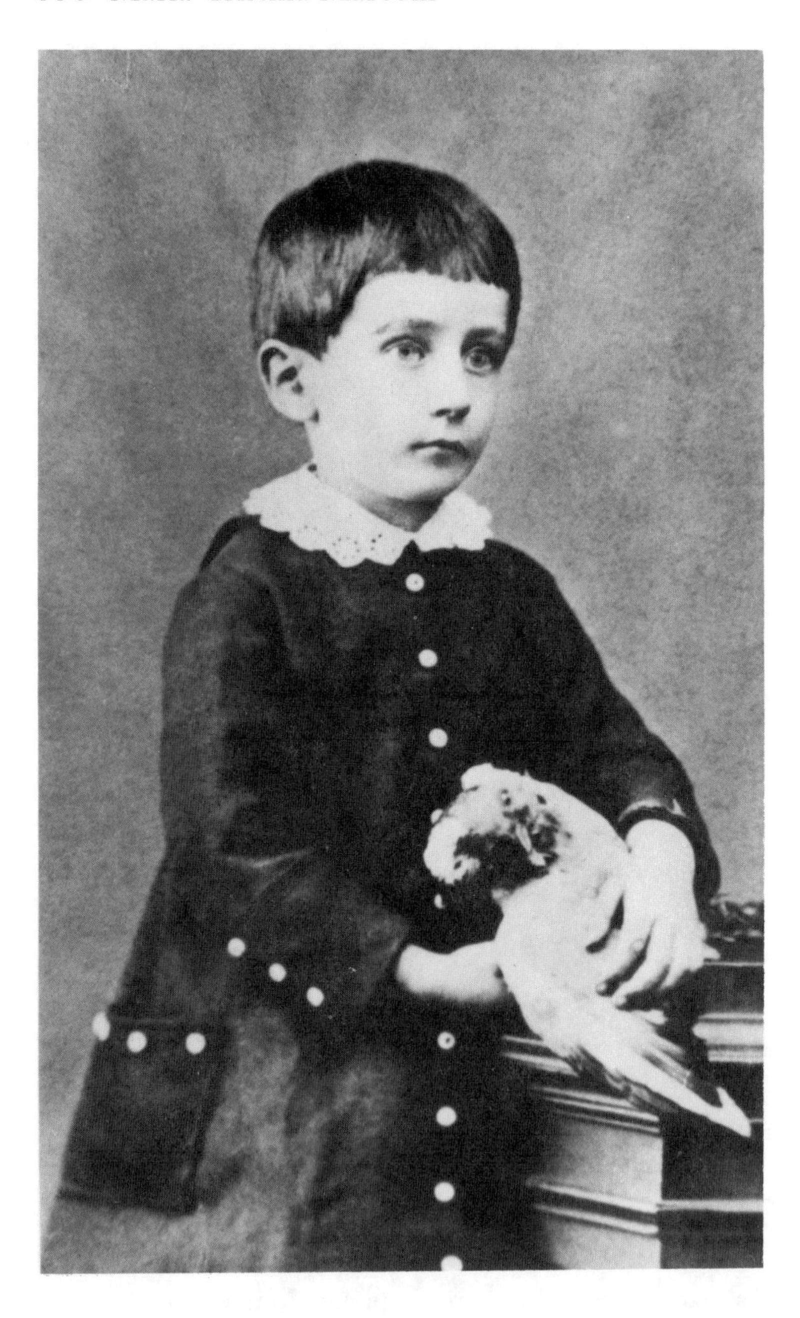

2. Guglielmo Marconi at age four.

3. Villa Griffone where, at age twenty one (1895), Marconi made his first invention on wireless communication.

4. Marconi (center) with his family, seated to the left, his father Giuseppe Marconi, his mother Annie Jameson, to his right, his oldest brother Alfonso, standing. Guglielmo gloomy look was due to the fact that Italians in general had ignored his invention.

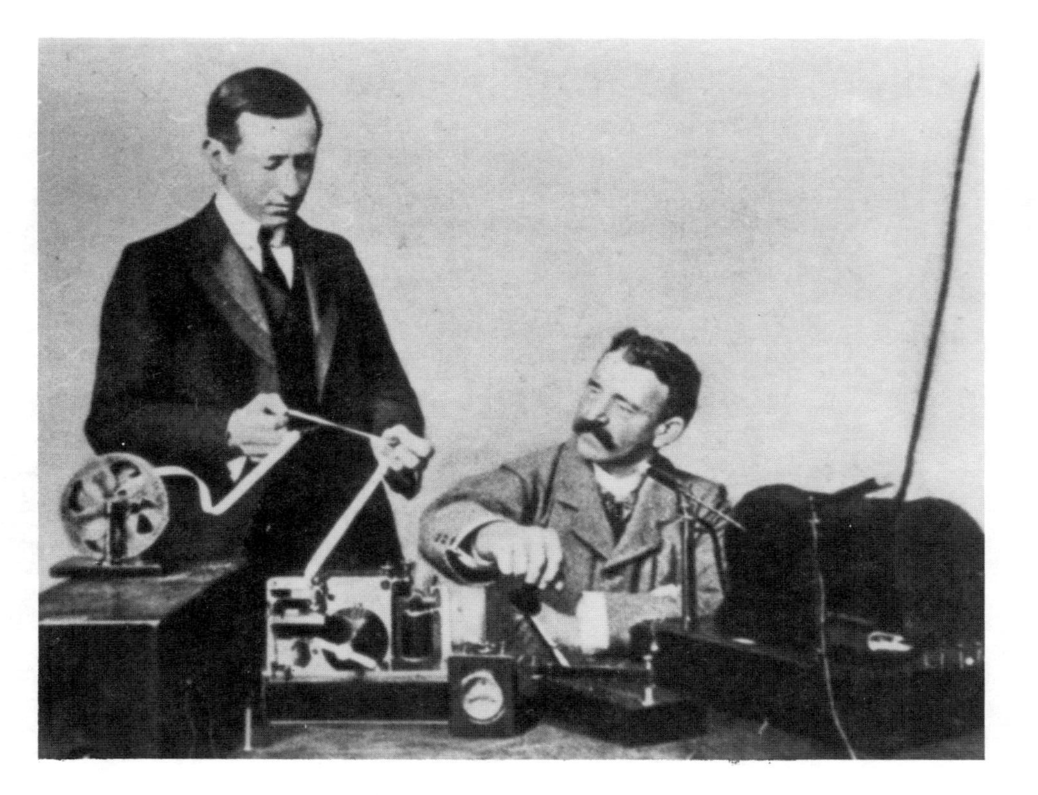

5. In England, Marconi with his faithfull assistant, George Kemp, 1896.

6. Poldhu Hotel and Marconi's Radio Station. Marconi chose the site because of its location on the Atlantic Ocean and because of the beaty of its surroundings.

7. St. John's Newfoundland: Marconi (extreme left) supervising the flight of the kite, with its attached antenna.

8. Glace Bay, Canada: Marconi, surrounded by a team of devoted engineers, during the construction of the Radio Station.

9. Poldhu: Royal Party of Prince and Princess of Whales visiting Marconi's Radio Station--18 July 1903.

10. Marconi, just 29 years old, at the height of his international successful career as an inventor--autographed picture.

11. Marconi receiving his Knighthood from King George V of Great Britain.

12. Mayor Angelo J. Rossi welcomes Marconi and his wife Maria Cristina--San Francisco, 1 October 1933.

13. Marconi with his wife, Marchesa Maria Cristina, in Hollywood with George Burns and friend (name unknown).

14. Marconi with Maria Cristina posing for an official photograph with passengers aboard the *Chichibumarú*, a prestigeous modern ship of Japan.

15. Marconi with his wife Maria Cristina on their stay in Japan, posing with their elegant Japanese friends. The Japanese lady is Yori Takahashi, who became a great friend of Maria Cristina.

16. Marconi and his wife Maria Cristina with Italian Ambassador to Japan, Giacinto Auriti (to the right), visiting the enormous Buddha of Kamakura.

17. Marconi with his wife Maria Cristina and friends in Singapore, aboard the luxury Italian liner, *Conte Rosso*.

18. Marconi with Maria Cristina aboard the *Conte Rosso*, surrounded by welcoming dignitaries of Bombay, India.

19. The daughter, Elettra, prefers to sit on her father's lap and watch rather than participate in the various field track events. Looking on are Maria Cristina and friend.

20. Last photograph of Marconi with his daughter Elettra, before his sudden death--20 July 1937.

# APPENDIX

# HONORS AND DECORATIONS CONFERRED ON GUGLIELMO MARCONI DURING HIS LIFETIME

1897 Knighthood conferred by the Italian Government.
1902 Knight of the Order of St. Anne (Russia).
1905 Knight of the Civil Order of Savoy (Italy).
**1909 Nobel Prize for Physics**
1912 Grand Cross of the Order of Alfonso XII. Spain.
1912 Cross of the Order of the Crown of Italy
1914 Knight Grand Cross of the Victorian Order (Great Britain).
1914 Tribute of the British Council of the "Royal Society of Arts" (Great Britain).
1914 Senator of the Realm of Italy.
1919 Iron Cross for military valour for services rendered during the First World War.
1927 Silver Medal of the "International Mark Twain Society" (USA).
1927 President of the National Research Council (CNR, Italy).
1929 Knight of the Plus Ultra Order (Spain).
1929 Title of Marchese conferred by King Vittorio Emanuele III.
1930 President of the Royal Academy of Italy.
1932 "Lord Kelvin" Medal of the "Institute of Civil Engineers" (Great Britain).
1932 "John Scott" Medal of the City of Philadelphia (USA).
1932 "Goethe" Medal from the President Paul von Hindenburg (Germany).
1933 Grand Cordon of the Order of the Rising Sun (Japan).

1935 Grand Cordon of the Order of the Southern Cross (Brazil).
1936 Grand Cross of the Order of Jade (China).
1936 Rear Admiral of the Italian Royal Navy Reserve.
1936 Grand Cross of the Order of Menelik (Ethiopia).

FURTHER HONOURS
Grand Cross of the "Ordine Piano" (Vatican City)
Grand Cross of the Sovereign Military Order of Malta
Grand Cross of the Order of St. Maurice and St. Lazarus (Italy)
Gold medal of the "Institute of Radio Engineers", New York (USA).
Tribute from the "Franklin Institute", Philadelphia (USA).
Gold plaque presented by the survivors of the tragic sinking of the Titanic (1912).
Honorary Degree from the University of Cambridge (Great Britain)
Honorary Degree from the University of Cambridge (Great Britain)
Honorary Degree from the University of Glasgow (Scotland)
Honorary Degree from the University of Liverpool (Great Britain)
Honorary Degree from the University of Pennsylvania (USA)
Honorary Degree from Columbia University (USA)
Honorary Degree from the Catholic University of Notre-Dame (France)
Honorary Degree from the University of Loyola
Honorary Degree from the Northwestern University of Chicago (USA)
Honorary Degree from the University of Bologna (Italy)
Honorary Degree from the University of Pisa (Italy)
Honorary Degree from the University of Rio de Janeiro (Brazil)
Honorary Degree from the University of Santa Clara
Honorary Degree from Stanford University

Plaque to Marconi (Portuguese)

A GUGLIELMO MARCONI

CIENTISTA ITALIANO
CIDADÃO DO MUNDO
QUE DO ENGENHO HUMANO
AFIRMOU O PERENE VALOR,
A VITÓRIA
SOBRE AS DISTÂNCIAS
E ILUMINOU DE ROMA, CIDADE ETERNA,
ESTE SÍMBOLO DO RIO, CIDADE MARAVILHOSA.

A SOC. ITALIANA DE BENEFICÊNCIA
E A COLETIVIDADE ITALIANA DO RIO
NO 1º CENTENÁRIO DO NASCIMENTO
COM REVERENTE MEMÓRIA OFERECEM
25 DE JULHO DE 1974

Plaque to Marconi (Italian)

## U.S. Patent Office, Transmitting Electrical Signals

# UNITED STATES PATENT OFFICE.

GUGLIELMO MARCONI, OF LONDON, ENGLAND.

## TRANSMITTING ELECTRICAL SIGNALS.

**SPECIFICATION forming part of Letters Patent No. 586,193, dated July 13, 1897.**

Application filed December 7, 1896. Serial No. 614,833. (No model.)

*To all whom it may concern:*

Be it known that I, GUGLIELMO MARCONI, student, a subject of the King of Italy, residing at 21 Burlington Road, London, in the county of Middlesex, England, have invented certain new and useful Improvements in Transmitting Electrical Impulses and Signals and in Apparatus Therefor, of which the following is a specification.

According to this invention electrical signals, actions, or manifestations are transmitted through the air, earth, or water by means of oscillations of high frequency, such as have been called the "Hertz rays" or "Hertz oscillations." Usually all line-wires are dispensed with. At the transmitting-station I employ a Ruhmkorff coil, having in its primary circuit a Morse key or other signaling instrument and at its poles appliances for producing the desired oscillations. The Ruhmkorff coil may, however, be replaced by any other source of high-tension electricity. When working with large amounts of energy, it is, however, better to keep the coil or transformer constantly working for the time during which one is transmitting, and instead of interrupting the current of the primary interrupting the discharge of the secondary. In this case the contacts of the key should be immersed in oil, as otherwise, owing to the length of the spark, the current will continue to pass after the contacts have been separated. At the receiving-station there is a local-battery circuit, containing any ordinary receiving instrument and an appliance for closing the circuit, the latter being actuated by the oscillations from the transmitting-station. When transmitting through the air and it is desired that the signal should only be sent in one direction, I place the oscillation-producer at the transmitting-station in the focus or focal line of a reflector directed to the receiving-station, and I place the circuit-closer at the receiving-station in a similar reflector directed toward the transmitting-station. When transmitting signals through the earth, I connect one end of the oscillation-producer and one end of the circuit-closer to earth and the other ends to similar plates, preferably electrically tuned with each other in the air and insulated from earth.

Figure 1 is a diagrammatic front elevation of the instruments at the transmitting-station when signaling through the air, and Fig. 2 is a vertical section of the transmitter. Fig. 2ª is a longitudinal section of the oscillator to a larger scale. Fig. 3 shows a detail on a larger scale. Fig. 4 is a diagrammatic front elevation of the instruments at the receiving-station. Fig. 5 is a full-sized view of the receiver. Fig. 6 shows a modification of the tube *j*. Fig. 7 shows the detector. Fig. 8 is a full-sized view of the liquid-resistance. Figs. 9 and 10 show modifications of the arrangements at the transmitting-station. Fig. 11 shows a modification of the arrangements at the receiving-station.

Referring now to Fig. 1, *a* is a battery, and *b* an ordinary Morse key closing the circuit through the primary of a Ruhmkorff coil *c*. The terminals $c'$ of the secondary circuit of the coil are connected to two metallic balls *d d*, fixed by heat or otherwise at the ends of tubes $d'$ $d'$, Fig. 2ª, of insulating material, such as ebonite or vulcanite. *e e* are similar balls fixed in the other ends of the tubes $d'$. The tubes $d'$ fit tightly in a similar tube $d^2$, having covers $d^3$, through which pass rods $d^4$, connecting the balls *d* to the conductors. One (or both) of the rods $d^4$ is connected to the ball *d* by a ball-and-socket joint and has a screw-thread upon it working in a nut in the cover $d^3$. By turning the rod therefore the distance of the balls *e* apart can be adjusted. $d^5$ are holes in the tube $d^2$, through which vaseline, oil, or like material is introduced into the space between the balls *e*

The balls *d* and *e* are preferably of solid brass or copper, and the distance they should be apart depends on the quantity and electromotive force of the electricity employed, the effect increasing with the distance so long as the discharge passes freely. With a coil giving an ordinary eight-inch spark the distance between *e* and *e* should be from one twenty-fifth to one-thirtieth of an inch and the distance between *d* and *e* about one and a half inches. *f* is a cylindrical parabolic reflector made by bending a metallic sheet, preferably of brass or copper, to form and fixing it to metallic or wooden ribs $f'$. Other conditions being equal the larger the balls the greater is the distance at which it is possible to communicate. I have generally used

# Schematic for Marconi's receiver for electrical oscillations

No. 668,315. Patented Feb. 19, 1901.

G. MARCONI.

RECEIVER FOR ELECTRICAL OSCILLATIONS.

(Application filed July 17, 1900.)

(No Model.)

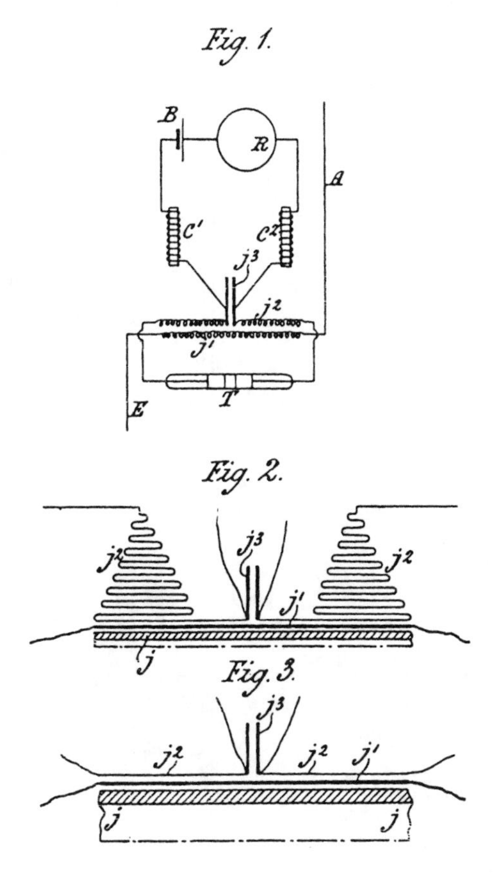


Witnesses.
E. A. Pollock
A. P. Hollingsworth

Inventor.
Guglielmo Marconi
By his Attorneys,
Baldwin Davidson & Wight.

(No Model.) 3 Sheets—Sheet 1.

G. MARCONI.

TRANSMITTING ELECTRICAL SIGNALS

No. 586,193. Patented July 13, 1897.

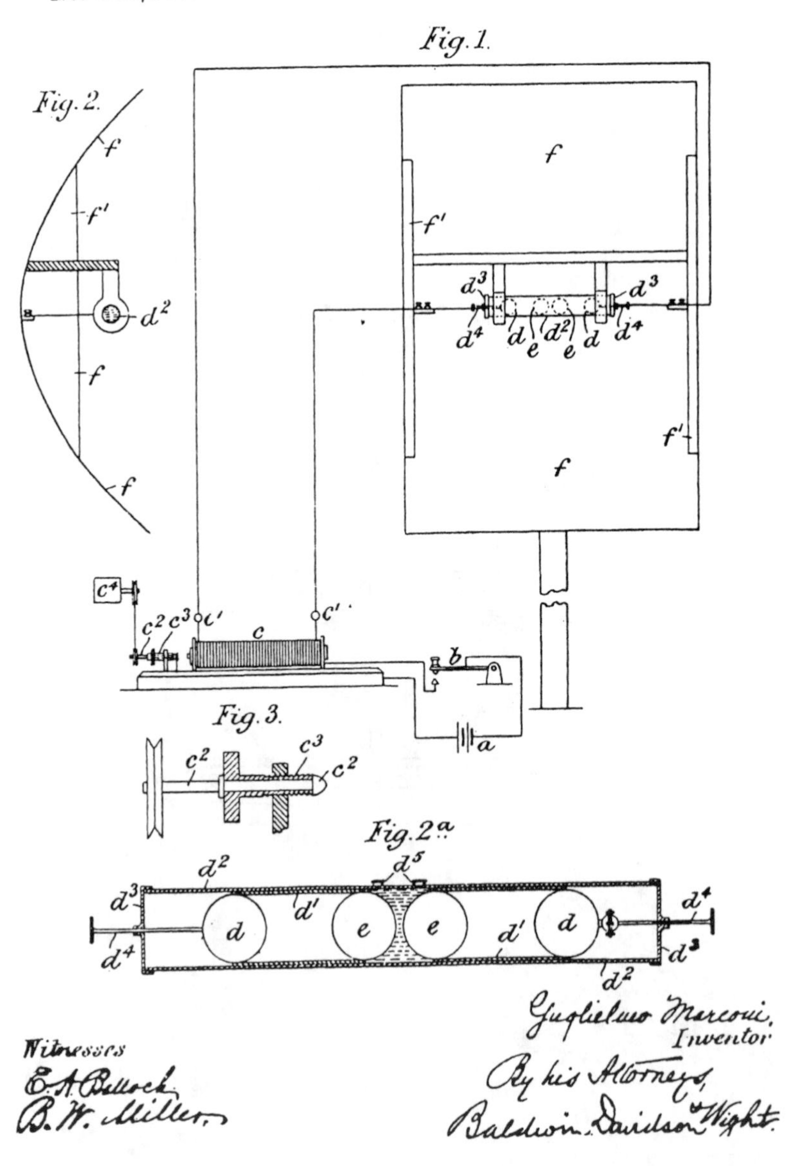

# DOCUMENTI D'EPOCA
# RECENT ADVANCES IN WIRELESS TELEGRAPHY.[a]

By Chevalier G. Marconi, LL. D., D. Sc., M. R. I.

The phenomena of electro-magnetic induction, revealed chiefly by the memorable researches and discoveries of Faraday carried out in the Royal Institution, have long since shown how it is possible for the transmission of electrical energy to take place across a small air space between a conductor traversed by a variable current and another conductor placed near it, and how such transmission may be detected and observed at distances greater or less, according to the

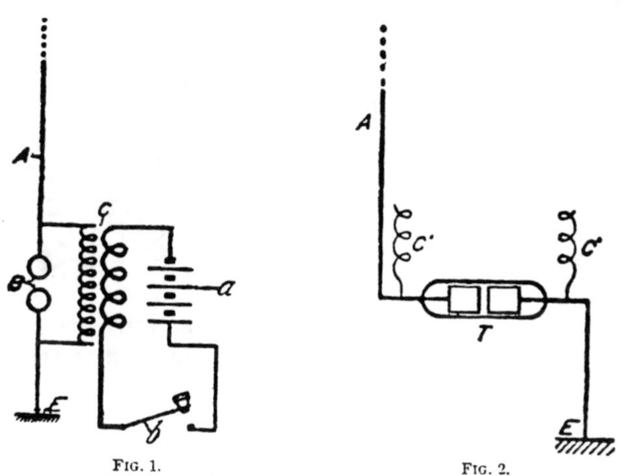


Fig. 1. Fig. 2.

more or less rapid variation of the current in one of the wires, and also according to the greater or less quantity of electricity brought into play.

Maxwell, inspired by Faraday's work, gave to the world in 1873 his wonderful mathematical theory of electricity and magnetism, demonstrating on theoretical grounds the existence of electro-magnetic waves, fundamentally similar to but enormously longer than

[a] Abstract of paper read before the Royal Institution of Great Britain at its weekly evening meeting, Friday, March 3, 1905. Reprint of extract from transactions of the Royal Institution.

tion going on through it overheard, or its operation interfered with. Sir William Preece has published results which go to show that it is possible to pick up at a distance on another circuit the conversation which may be passing through a telephone or telegraph wire.

Up to the commencement of 1902 the only receivers that could be practically employed for the purposes of wireless telegraphy were based on what may be called the coherer principle—that is, the detector, the principle of which is based on the discoveries and observations made by S. A. Varley, Professor Hughes, Calsecchi Onesti, and Professor Branly.

Early in that year the author was fortunate enough to succeed in constructing a practical receiver of electric waves, based on a principle different from that of the coherer. Speaking from the experience of its application for over two years to commercial purposes, the author is able to say that, in so far as concerns speed of working, facility of adjustment, reliability, and efficiency when used on tuned circuits, this receiver has left all coherers or anticoherers far behind.

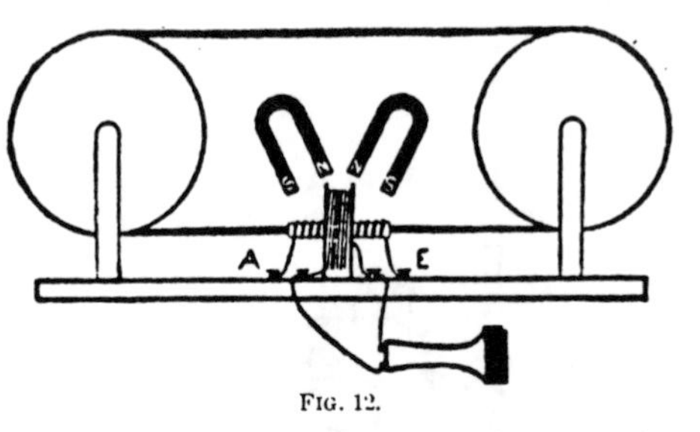

FIG. 12.

The action of this receiver is in the author's opinion based upon the decrease of magnetic hysteresis, which takes place in iron when under certain conditions this metal is exposed to high frequency oscillations of Hertzian waves.

It is constructed in the following manner and is shown in fig. 12.

On an insulating sleeve surrounding a portion of a core, consisting of an endless rope of thin iron wires, are wound one or two layers of thin insulated copper wires. Over this winding insulating material is placed, and over this again another longer winding of thin copper wire contained in a narrow bobbin. The ends of the windings nearer the iron core are connected one to earth and the other to the elevated conductor, or they may be joined to any suitable syntonizing circuit, such as is now employed for syntonic wireless telegraphy. The ends of the longer winding are connected to the terminals of a suitable telephone. A pair of horseshoe magnets are conveniently disposed for magnetizing the portion of the core surrounded by the windings, and the endless iron core is caused to move continuously through the windings and the field of the horseshoe magnets.

This detector is and has been successfully employed for both long and short distance work. It is used on the ships of the Royal Navy and on all trans-Atlantic liners which are carrying on a long-distance

It was possible nearly five years ago to send different messages simultaneously without interference, the messages being received on differently tuned receivers connected to the same vertical conductor.

This result was described in the Times of October 4, 1900, by Professor Fleming, who, in company with others, witnessed the test.

A recent improvement introduced in the method of tuning the receiver is that shown in fig. 11.

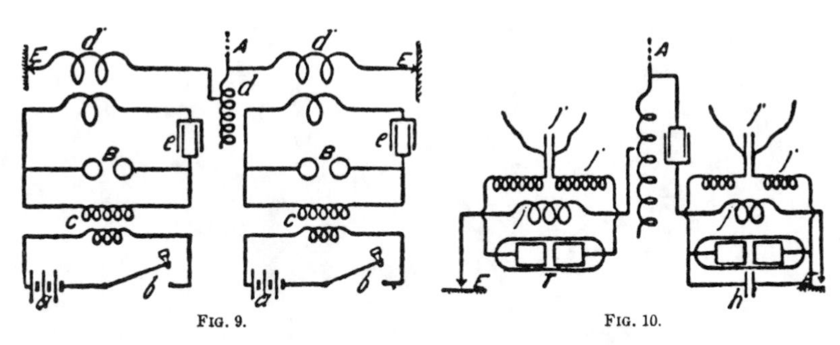


FIG. 9. FIG. 10.

There exists at present among the large section of the public considerable misconception as to the feasibility of tuning or syntonizing wireless telegraphic installations, and also as to what is generally termed "the interception of messages." According to the accepted understanding, "intercepting" a message means or implies securing by force, or by other means, a communication which is intended for somebody else, thereby preventing the intended recipient from receiving it. Now, this is just what has never happened in the case of wireless telegraphy. It is quite true that messages are, and have been, tapped or overheard at stations for which they are not intended, but this does not by any means prevent the messages from reaching their proper destination. Of course, if a powerful transmitter giving off strong waves of different frequencies is actuated near one of the receiving stations, it may prevent the reception of messages, but the party working the so-called interfering station is at the same time unable to read the message he is trying to destroy, and therefore the message is not, in the popular sense of the word, "intercepted." It should be remembered that any telegraph or telephone wire can be tapped, or the conversa-

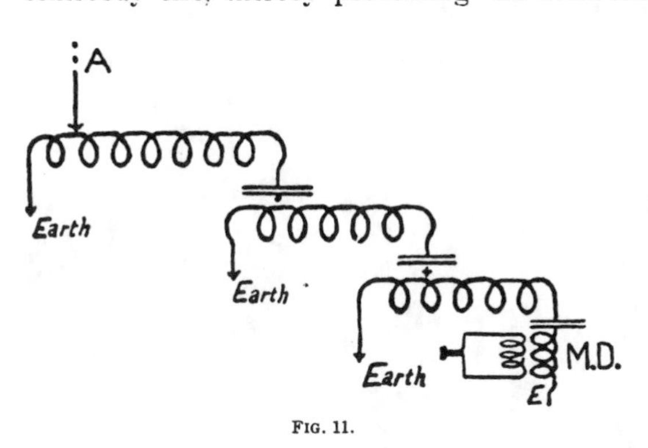


FIG. 11.

Marconi in front of his newly invented characteristic parabolic antenna, used in his experiments with short waves, from 1931 to 1934 in the Gulf of Tigullio.

## Marriage Certificate of Guglielmo Marconi and Maria Cristina

N° 285

Marconi e Bezzi-Scali

L'anno millenovecento ventisette nel giorno quindici (15) del mese di Giugno pubblicazioni nel tempo della celebrazione della S. Messa ........ chiesa parrocchiale d ........ e ........ in quella d ........ con dispensa ........

e non scopertosi verun canonico impedimento S. E. Rev.ma il Sig.r Cardinale Lucidi Evaristo qui sottoscritto ........ ha congiunto in matrimonio secondo il rito di S. Madre Chiesa il Sig.r Senatore Marconi Guglielmo nato in ........ domiciliato ........ parrocchia d ........

Documenti

e la Sig.na Contessa Bezzi Scali Maria Cristina nata in ........ domiciliata nella parrocchia di S. Andrea delle Fratte

Testimoni S. E. il Governatore Lodovico Principe Spada Potenziani fu Giovanni, da Rieti, Governatore di Roma; S. E. Clemente Del Drago di Luigi, Romano, parr. S. Cuore; Sig.r Principe Barberini Luigi fu Urbano Sacchetti, Romano, parr. S. Camillo; Sig.r Marchese Guglielmi Guglielmo fu Giuseppe, Romano parr. S. Carlo [illegible]

Firma degli Sposi: Guglielmo Marconi; Maria Cristina Bezzi Scali Marconi [illegible]

Firma del Parroco: Card. Evaristo Lucidi

Firma dei Testimoni: Lodovico Spada Potenziani; Clemente Del Drago; Luigi Barberini; Guglielmo Guglielmi

Facsimile of Marchesa Maria Cristina's letter to Cav. Peoni on behalf of the *Eletra*

11 Via Condotti Roma
16 ottobre 1977.

Gentilmo Cav. Peoni

Di ritorno a Roma nella mia corrispondenza trovo la sua comunicazione del 30-8-u.s.

Prima di tutto desidero dirle quanto mi ha fatto piacere notare l'interessamento suo e dei Cav. Anziani della Marconi Italiana riguardo il tanto amato Yacht Elettra.

Desidero vivamente che lei sappia insieme alla Compagnia Marconi Inglese a Chelmsford quanto io personalmente mi sia prodigata per diversi anni con tutte le mie forze presso Ministri e le più alte personalità dello Stato Italiano per salvare lo yacht Elettra

Iniziai subito dopo il rilascio da parte di Tito dello yacht Elettra che si trovava allora a Zara e che in seguito fu rimorchiato a Trieste.

Durante gli anni che lo yacht Elettra giaceva a Muggia, Trieste, il pensiero e l'animo mio e di mia figlia Elettra, alla quale il Padre ha voluto dare il nome della nave in ricordo delle sue navigazioni per gli esperimenti soprattutto riguardo il „Beam System„ è stato sempre rivolto allo yacht Elettra con ansia e preoccupazione.

Il mio dolore è sempre vivo, anche perchè ho passato sullo Yacht Elettra parte della mia vita vicina al mio caro consorte.

Il governo Italiano ha sempre dimostrato molte difficoltà finanziarie per riabilitare lo Yacht Elettra che avrebbe mantenuto vivo il ricordo di Guglielmo Marconi anche sotto l'aspetto scientifico.

Sarei molto contenta se la Marconi Italiana potesse realizzare il suo desiderio di ottenere un cimelio della nave, avrei piacere se mi tenessero al corrente,

Intanto voglia gradire le mie migliori espressioni.

Maria Cristina Marconi

Cil. cav. Angelo Peoni
Presidente „Marconi Italiana„
ass. naz. lav. anziani
Via A. Negrone 1 -
Genova - Cornigliano -

English language translation of Marchesa's letter:

Letter autographed and written by Marchesa Maria Cristina Marconi--11 Via Condotti Roma--16th October 1977

Esteemed Cav. Peoni
Going back to Rome, in my mail I found your letter of 30-8-1977. First of all, I wish to tell you how much I have appreciated the interest that you show and even the Loyer Anzioni of the Marconi (Company) in Italy, regarding the beloved yacht Elettra.

I wish very much that you should know together with the Marconi Company in Chelmsford, England, how much I personally have worked for so many years with all my strength contacting the Ministers and the highest personalities of the Italian government to save the yacht Elettra.

I started immediately after Tito gave back the yacht Elettra which was then in Zara Harbor and towed to Trieste.

During the years the Elettra anchored in Muggia, Trieste, my thoughts and my spirit as well as that of my daughter, Elettra, to whom the Father had given the name of his yacht in remembrance of his sailing on board making his experiments regarding the "Beam System" has always been extended or directed to the yacht Elettra with anxiety and great worry.

My sorrow is always very deep, also because I spent part of my life near my beloved husband on board the Elettra.

The Italian government has always placed many financial obtacles to restore the yacht Elettra which would have kept alive the memory of Giglielmo Marconi also from the scientific point of view.

I would be very happy if the Marconi Company in Italy could realize his wish to obtain a relic of the yacht Elettra. Will you please (if you could) inform me about it.

With best regards
Maria Cristina Marconi

# INDEX